Intelligent Support for Computer Science Education

Intelligent Support for Computer Science Education

Pedagogy Enhanced by Artificial Intelligence

Barbara Di Eugenio

Davide Fossati

Nick Green

CRC Press
Taylor & Francis Group
Boca Raton London New York

CRC Press is an imprint of the
Taylor & Francis Group, an **informa** business

First edition published 2022
by CRC Press
6000 Broken Sound Parkway NW, Suite 300, Boca Raton, FL 33487-2742

and by CRC Press
2 Park Square, Milton Park, Abingdon, Oxon, OX14 4RN

CRC Press is an imprint of Taylor & Francis Group, LLC

Library of Congress Cataloging-in-Publication Data

Names: Di Eugenio, Barbara, author. | Fossati, Davide, author. | Green, Nick (Engineer), author.
Title: Intelligent support for computer science education : pedagogy enhanced by artificial intelligence / Barbara Di Eugenio, Davide Fossati, Nick Green.
Description: First edition. | Boca Raton : CRC Press, 2022. | Includes bibliographical references and index.
Identifiers: LCCN 2021013481 | ISBN 9781138052017 (hardback) | ISBN 9781032049861 (paperback) | ISBN 9781315168067 (ebook)
Subjects: LCSH: Computer science--Study and teaching--Data processing. | Artificial intelligence. | Computer-assisted instruction.
Classification: LCC QA76.27 .D49 2022 | DDC 004.071--dc23
LC record available at https://lccn.loc.gov/2021013481

ISBN: 978-1-138-05201-7 (hbk)
ISBN: 978-1-032-04986-1 (pbk)
ISBN: 978-1-315-16806-7 (ebk)

Typeset in CMR10 font
by KnowledgeWorks Global Ltd.

Contents

SECTION II From Human Tutoring to ChiQat-Tutor

WITH STELLAN OHLSSON, MEHRDAD ALIZADEH, LIN CHEN, and RACHEL HARSLEY

WITH OMAR ALZOUBI and CHRISTOPHER BROWN

Section III Extending ChiQat-Tutor

About the Authors

Barbara Di Eugenio is a Professor in the Department of Computer Science at the University of Illinois at Chicago (UIC), Chicago, IL, USA. There she leads the NLP laboratory (https://nlp.lab.uic.edu). Dr. Di Eugenio holds a Ph.D. in Computer Science from the University of Pennsylvania (1993); she joined UIC in 1999. Her interests focus on the theory and practice of Natural Language Processing, with applications to educational technology, health care, human robot interaction and social media. Dr. Di Eugenio is an NSF CAREER awardee (2002), and a UIC University Scholar (2018–21). Her research has been supported by the National Science Foundation, the National Institute of Health, the Office of Naval Research, Motorola, Yahoo!, Politecnico di Torino and the Qatar Research Foundation. She has graduated 12 Ph.D. students and 30 Master's students, and published more than one hundred refereed publications.

Davide Fossati (http://www.fossati.us) is currently a Senior Lecturer in Computer Science at Emory University in Atlanta, GA, USA. Prior to joining the faculty at Emory in 2016, Dr. Fossati held positions at the Georgia Institute of Technology (2009–2010) and Carnegie Mellon University (2010–2015). He received his Ph.D. in Computer Science from the University of Illinois at Chicago in 2009. He also holds an M.Sc. degree in Computer Engineering from the Politecnico di Milano, Italy (2004), and an M.Sc. in Computer Science from the University of Illinois at Chicago (2003). Dr. Fossati's primary scholarly focus is Technology Enhanced Learning, with particular interest in the development of Artificial Intelligence systems to support Computer Science education.

Nick Green (http://nickgreen.co) is a technology professional with 20 years of research and development experience in academia and industry. Dr. Green received his Ph.D. in Computer Science from the University of Illinois at Chicago in 2017, where he focused on educational technology, natural language processing and software engineering. Outside of academia, he has worked for companies such as Sony® Interactive Entertainment and Facebook®. He has a passion for the startup scene where he is also a serial entrepreneur having founded companies in fields such as security and precision agriculture.

Contributors

Mehrdad Alizadeh
Lexis Nexis
Raleigh, NC, USA

Omar AlZoubi
Jordan University of Science and Technology
Irbid, Jordan

Christopher Brown
United States Naval Academy
Annapolis, MD, USA

Lin Chen
Cambia Health Solutions
Seattle, WA, USA

Rachel Harsley
Google
Austin, TX, USA

Stellan Ohlsson
University of Illinois at Chicago
Chicago, IL, USA

Preface

This book presents our research journey into the effectiveness of human tutoring, with the goal of developing educational technology that can be used to improve introductory Computer Science education at the undergraduate level. Nowadays, Computer Science education is central to the concerns of society, as attested by the penetration of information technology in all aspects of our lives; consequently, in the last few years, interest in Computer Science at all levels of schooling, especially at the college level, has been burgeoning. However, introductory concepts in Computer Science such as data structures and recursion are difficult for novices to grasp, even for college students. At the same time, educational technology has emerged as a field of research in its own that can support education and instruction at scale; specifically, Intelligent Tutoring Systems attempt to replicate the efficacy of one-on-one human tutoring.

In our research, we explore the pedagogy of teaching those difficult introductory concepts in human–human, one-on-one tutoring dialogues; and we use our findings to inform interaction strategies in ChiQat-Tutor, the Intelligent Tutoring System we developed. In turn, these findings are not exclusive to Computer Science, but concern one-on-one tutoring more in general.

Our research journey was highly interdisciplinary from the start and is grounded in four distinct disciplines: cognitive theories of learning; linguistic theories of human conversation; the pedagogy of Computer Science and educational technology. One of the goals we pursue in this book is to uncover operational correlations that illuminate how learning mechanisms, as hypothesized in a specific theory of learning, are enacted in human–human interaction, and to define the conditions under which each mechanism is adopted by the participants, in our case, human tutors who finely tailor their responses to students' problem solving steps, questions and suggestions.

Next, we describe how we computationally modeled those pedagogical strategies in progressively more sophisticated versions of ChiQat-Tutor. In total, 11 different versions of ChiQat-Tutor were evaluated in controlled experiments in actual classrooms and laboratories. The results we describe were obtained at three very different institutions: a public, urban university, the University of Illinois at Chicago, Chicago, USA; the undergraduate college of the USA naval service, the United States Naval Academy, Annapolis, MD, USA and the Middle East campus of a private University, Carnegie Mellon University in Qatar, Doha, Qatar. Careful evaluation is needed because no matter how much principled effort is poured into an educational technology system, a vast gap remains between human educators and their computational counterparts, that can only embody a minimal part of the richness of a human being. Because of this gap, and because of the limitations of technology, it is not necessarily the case that what works between humans works between a learner and a computer system. At the same time, experimentation with a software system can pinpoint which specific aspects of interaction are more conducive to learning, precisely because of the more controlled number of parameters under analysis.

The book concludes with a practical guide to adapting ChiQat-Tutor to different topics and potentially domains. ChiQat-Tutor is available and downloadable from http://www.digitaltutor.net/, and so is the original annotated corpus, at http://www.digitaltutor.net/chiqat/pages/database.php.

The intended audience of this book is as broad as our foundations. To start with, this book will be of special interest to the Computer Science education community, specifically as concern introductory courses at the college level, and Advanced Placement (AP) courses at the high school level. In addition to our own work, we also provide a comprehensive and succinct overview of the Computer Science education landscape. Second, all our work is particularly relevant to the Educational Technology community, especially to those working in Intelligent Tutoring Systems, their interfaces and Educational Data Mining.

Besides these two fundamental constituencies, additional audiences include: cognitive psychologists interested in how human learning manifests itself in interaction; the broad Natural Language Processing community, including computational linguists, and researchers interested in dialogue analysis, and Human Computer Interaction; the broad Artificial Intelligence community, and in particular, practitioners of machine learning and data mining on small datasets; educators for sister disciplines to Computer Science, for example Information Technology.

Acknowledgments

This work started a long time ago, in the late 1990s, when Barbara Di Eugenio was a post-doctoral researcher with Prof. Johanna D. Moore at the Learning Research and Development Center of the University of Pittsburgh. Prof. Moore introduced Di Eugenio to Natural Language Processing for Educational Technology, and to Dr. Susan Chipman, the Office of Naval Research program officer who for many years strongly supported research in cognitive foundations of learning, and in language interfaces to educational technology, well before this was a fashionable topic in the mainstream.

Several years later, as a faculty member at the University of Illinois at Chicago, Di Eugenio introduced several generations of graduate students to this line of work, including Davide Fossati and Nick Green, co-authors of this book. The collaboration with former students – now faculty members or industry professionals themselves – made it possible for this project to cross the boundaries of academic institutions and even countries.

A research program that spans almost 15 years benefits from contributions from a vast group of collaborators, some who were directly involved in the project, and some who were part of the intellectual milieu which makes research possible.

Prof. Stellan Ohlsson (Psychology, University of Illinois at Chicago) significantly contributed to our study of human tutoring, and contributed to Chapters 2, 3 and 5 of this book. Prof. Christopher Brown (Computer Science, United States Naval Academy) worked closely with us in the development and evaluation of several initial versions of ChiQat-Tutor, and contributed to Chapter 4 of this book. Prof. Omar AlZoubi (at the time post-doctoral researcher at Carnegie Mellon University in Qatar, currently faculty member at the Jordan University of Science and Technology) contributed to the development and evaluation of later modules of ChiQat-Tutor, and contributed to Chapters 4 and 6 of this book.

We also extend our gratitude to the following faculty members, who contributed in various capacities such as tutors, advisors and facilitators for our evaluation experiments in their courses: John Bell (UIC), Tanya Berger-Wolf (UIC), Ugo Buy (UIC), Lowell Carmony (Lake Forest College), Bhaskar DasGupta (UIC), Martha Evens (Illinois Institute of Technology), Susan Goldman (UIC), Mark Guzdial (University of Michigan), John Lillis (UIC), Agathe Merceron (Beuth Hochschule für Technik, Berlin, Germany), Tom Moher (UIC), Don Needham (United States Naval Academy), Dale Reed (UIC), Patrick

Seeling (Central Michigan University), Mitchell Theys (UIC) and Patrick Troy (UIC).

Many graduate students directly contributed to this work, helping in a variety of tasks including (but not limited to) annotating and analyzing data, running experiments, grading tests, developing software and writing documents. From Computer Science, UIC: Jisha Abubaker, Sabita Acharya, Mehrdad Alizadeh (contributor to Chapters 3 and 6), Lin Chen (contributor to Chapter 3), Rachel Harsley (contributor to Chapters 3 and 5), Cindy Howard (Kersey), Xin Lu, Camillo Lugaresi, Zhuli Jack Xie. From Psychology, UIC: Bettina Chow, David Cosejo, Andrew Corrigan-Halpern, Trina C. Kershaw, Justin Oesterreich.

Additionally, the following graduate students contributed to a stimulating intellectual environment at UIC, and provided valuable advice: Anushka Anand, Joel Booth, Itika Gupta, Nitin Jindal, Abhinav Kumar, Paul Landes, Natawut Monaikul, David Randolph, Saad Sheikh, Rajen Subba and Alberto Tretti.

Several undergraduate students at the University of Illinois at Chicago (UIC) and Carnegie Mellon University in Qatar (CMU-Q) also worked on this project, performing important tasks including tutoring, grading, transcribing and annotating data: Jack Bolous (UIC), Gibel Buena (UIC), Aidan Feldman (UIC), Sarah Jeziorski (UIC), Hira Dhamyal (CMU-Q), Sharjeel Khan (CMU-Q), Arsal Malik (CMU-Q) and Shahan Memon (CMU-Q).

This work was financially supported by multiple agencies and institutions over the years. The main sponsors of the research presented in this book were the Office of Naval Research (N00014–00–1–0640 and N00014-07-1-0040), the Qatar National Research Fund (NPRP 5–939–1–155) and the Graduate College of the University of Illinois at Chicago (2008/2009 Dean's Scholar Award to Fossati). Additional partial support came from the National Science Foundation (IIS–0133123, ALT–0536968, and CIF–0937060 which supported Fossati via a Computing Innovation Fellowship), and from a 2018/2021 UIC University Scholar Award to Di Eugenio.

It goes without saying that beyond the professional sphere, many other people contributed to this book, from near and far.

Barbara Di Eugenio would like to thank her husband Miloš Žefran and children Luca and Mattia; back in Italy, her mother Adelaide, sister Cristina, brother-in-law Alessandro, nephew Jacopo and many aunts and cousins and her mother-in-law Nada and brother-in-law Rastko and his family in Slovenia. Not only is Miloš Žefran an extremely supportive spouse but also he doubles up as a colleague, research collaborator and sounding board for new ideas.

Davide Fossati would like to thank his very supportive family: his father Carlo, who transmitted to him the passion for scientific inquiry; his mother Mariella, from whom he inherited the love for teaching; his brother Emanuele, who introduced him to the world of Computer Science; his aunts Lidia, Franca, Giulia, Anna; his uncle Faliero; his cousins Micaela and Alberico; his grandparents Alma, Alfiero, Egle and, last but not least, his partner Semo.

Nick Green would like to thank his wife, Nicole, for all the support she gave while he worked on his PhD thesis, and beyond. He would also like to thank his mentors during his graduate studies, Barbara Di Eugenio, and Ugo Buy, for giving him the chance to grow in the world of academia. To his parents, Irene and Jeff, for giving him a supportive childhood where anything was possible. Finally, to his daughter, Natalie for making a brighter future.

I

Four Scientific Pillars

Introduction

CONTENTS

T HE field of education goes back thousands of years, to ancient Egypt and Greece, and has continuously evolved over the millennia. In the last sixty years, a new force has come along to help shape education: information technology. This has never been more acutely felt than in the year 2020: because of the COVID-19 pandemic, technology infrastructure is supporting remote learning to an extent never seen before, and at the same time, it is causing a rethinking of pedagogical strategies more aligned with the explosion of information on the Internet and the myriad of available educational platforms. Recent developments in Artificial Intelligence and Machine Learning have opened up the promise of *radically [improving] human learning and education in formal settings as well as informal settings*, and support a grand challenge of "Education for All" *to radically expand access of learning to all Americans*, as recently stated by the National Science Foundation (NSF).[1] The NSF challenge is a contemporary version of earlier calls for *Advance Personalized Learning* or *Learning for Life* coming from various national research institutions [224], such as, respectively, the USA National Academy of Engineering and the United Kingdom Computing Research Committee (UKCRC). The work discussed in this book presents a stepping stone, in the specific discipline of Computer Science, on how to achieve these goals.

1.1 AN INTERDISCIPLINARY PERSPECTIVE

The impact of Artificial Intelligence on Education started about 50–60 years ago, with Computer-Aided Instruction (CAI) [56], which then morphed into the design and development of a variety of technologies to support learners and teachers, in both formal and informal settings. A non-exhaustive list of such technologies includes Intelligent Tutoring Systems, educational games,

---

[1] https://www.nsf.gov/pubs/2020/nsf20604/nsf20604.htm, visited Nov. 2020. The National Science Foundation (NSF) is the federal institution devoted to supporting research in the sciences, including social sciences, and engineering in the USA.

social robots and virtual agents deployed in educational scenarios. What unifies all these advances is the inherent interdisciplinarity of the field of Artificial Intelligence in Education, which is grounded in a variety of fields of inquiry beyond the two in its namesake, Education and Artificial Intelligence: from Psychology to Linguistics, from Social Sciences to Human-Computer Interaction. The flow of inquiry is not just one-way, from more theoretical disciplines such as Psychology to educational technology: the findings from computational modeling and user evaluation feed back to theories of learning, since they are one powerful way of assessing these theories. In recent years, the two-way flow of inquiry has been strengthened by the emergence of the field of Educational Data Mining (EDM): through EDM, insights from educational data can flow backward to theories of learning and forward to the development of educational technologies.

Our work is situated in this landscape of interdisciplinary research. Our specific, applied goal is to improve introductory Computer Science education at the undergraduate level, but this applied goal is contingent on several other scientific goals that we articulate below. Our methodology draws on four distinct disciplines: cognitive theories of learning, linguistic theories of human conversation, the pedagogy of Computer Science, and educational technology, in particular Intelligent Tutoring Systems (ITSs).

Two of our foundational premises are that pedagogical strategies and procedures can be made more effective by grounding them in a theory of learning, and that individual cognition is shaped by social interactions. Such social interactions are varied, happen both in formal and informal settings, and occur between learner and educator, among learners themselves and even in settings in which roles are more fluid, like visitors collaboratively interacting with exhibits in a museum. We specifically focus on one-on-one communication between student and teacher. Hence, one of our goals is finding operational correlates that illuminate how learning mechanisms, as hypothesized in a specific theory of learning are enacted in human–human interaction, and to define the conditions under which each mechanism is adopted by the participants, in our case specifically human tutors. Our findings on the functions of positive feedback and worked-out examples, for instance, transcend Computer Science as a specific application domain, and shed light on how human tutors modulate the pedagogical strategies they employ when interacting with their tutees.

We then aim to translate the findings from our analysis of one-on-one tutoring interactions into design principles for educational technology. There is often a gap between those results and their implementation in a working system, because the shape of those findings is often expressed as static correlations but not as a computational process; this is where computational modeling comes into the picture and results in a working software system, our ITS ChiQat-Tutor.

Finally, we experiment with the technology we developed as just described, in a valid educational setting, to assess the actual impact of the findings revealed by the analysis of human tutoring data, as rendered in ChiQat-Tutor. No matter how much principled effort is poured into an educational technology system, a vast gap remains between human educators and their computational counterparts, that can only embody a minimal part of the richness of a human being. Because of this gap, and because of the limitations of technology, it is not necessarily the case that what works between humans works between a learner and a computer system. At the same time, experimentation with a software system can pinpoint which specific aspects of interaction are more conducive to learning, precisely because of the limited number of parameters under analysis.

Going back to our applied goal of improving introductory Computer Science (CS) education, when we started our work in the mid-2000s, this choice was due to our interest as educators in this field, to the dearth of educational technology solutions in this space and to a convergence of interests between us and our cognitive psychology collaborator and contributor to several chapters in this book, Stellan Ohlsson. At the time, CS programs were vastly undersubscribed, following the *dot-com bubble* market crash of 2002.[2] The landscape could not be more different today.

CS and information technology have penetrated all aspects of modern society, and therefore, there is a push for *Computer Science for all*, in that every citizen should have enough insight into technology so as to make informed choices. Specifically at the college level, CS and information technology open paths to secure, well-remunerated and innovative employment that can have true impact on society; not surprisingly, for the last few years, CS enrollment has skyrocketed with double digit percent increases in Bachelor's degrees from 2013 to 2018 [444]. Even for the 2020–21 academic year when University enrollment has decreased because of the COVID-19 pandemic – freshman enrollment has declined by 13% according to the latest report by the Chronicle of Higher Education[3] – CS departments are among the few which have enjoyed levels of enrollments similar to previous years, at least as judged from reports provided by individual institutions, including Di Eugenio's UIC.[4]

Today, our goal of improving CS undergraduate education could not be more timely: the fundamental concepts in CS have not changed, even if as we will discuss in Section 2.3, the field is undergoing a transformation as concerns becoming more accessible to a broader and diverse swath of the population. Our approach demonstrates a rigorous dive into three specific

---

[2]https://www.theguardian.com/technology/2005/mar/10/newmedia.media. Accessed Dec. 2020.

[3]https://www.insidehighered.com/news/2020/11/12/enrollment-declines-continue-national-student-clearinghouse-finds. Accessed Dec. 2020.

[4]No general articles on freshman enrollment in CS programs in Fall 2020 are available as of December 2020, neither by the Chronicles of Higher Education, nor by the relevant professional societies: the Association for Computing Machinery (ACM) or the CRA (Computing Research Association).

data structures, and one algorithmic strategy, recursion, how they are tutored, how those strategies can be deployed in an ITS and what the evaluation with students enrolled in those classes reveals.

All the work we discuss in this book was completed by 2018, and hence, the last computational models date back to 2017; this means that the machine learning/data mining components we employed predate the "deep learning" revolution in Artificial Intelligence, i.e., the advent of the massive neural network models such as Generative Pre-trained Transformer 3 (GPT3), that every day appears to be more and more capable.[5] Even if we were to start our work today, we would be very cautious in our usage of these Machine Learning methods: they require huge amounts of data to train and enormous computational power, they are not explanatory, they face serious issues of bias and unfairness and some are not open source [38, 277, 395]. Indeed, deep learning has not overtaken the AI in Education field so far, because deep learning models are often black boxes with no explanatory power [363]: even when they perform better on say, a task such as assigning grades, they are not able to provide the reasons behind the specific score, which are needed to formulate feedback for the learner [431, 441]. In the conclusions to this book, we will come back to the issue of applying state-of-the art Deep Learning models to educational data.

1.2 THE STRUCTURE OF THE BOOK

The book is structured as follows. The first part, *The Four Pillars*, provides the background for our work. Chapter 2 gives an overview of our four foundations: cognitive theories of learning; linguistics, especially pragmatics and dialog processing; the history and pedagogy of introductory CS education and educational technology, specifically, Intelligent Tutoring Systems (ITSs). In each field's overview, after some general introduction, we hone in on the components that are most relevant to our work: Ohlsson's theory of Modes of Learning; theories of speech acts and computational models of dialogues; introductory CS education at the college level and ITSs that operate within introductory CS and interact with their users in natural language.

The second part, *From Human Tutoring to ChiQat-Tutor*, covers the main body of our work. Chapter 3 describes our human–human tutoring data and the in-depth analyses of that data that we conducted, grounded in the cognitive theory of learning we had adopted. Chapter 4 illustrates the computational design of ChiQat-Tutor, including its conceptual architecture with all its components: it shows how the findings from Chapter 3 were operationalized, and shows how we mined the logs collected in the first version of ChiQat-Tutor, ChiQat-Tutor-v1, to automatically infer the Procedural Knowledge Model that ChiQat-Tutor-v2 uses as the foundation for its dynamic student model.

---

[5]https://www.nytimes.com/2020/11/24/science/artificial-intelligence-ai-gpt3.html, https://beta.openai.com/. Accessed Dec. 2020.

Chapter 5 is devoted to the evaluation of ChiQat-Tutor-v1 and ChiQat-Tutor-v2, and all the variations thereof, with more than 530 students in classroom and laboratory sessions at two different institutions: UIC, the University of Illinois at Chicago (Chicago, IL) for both ChiQat-Tutor-v1 and ChiQat-Tutor-v2; and USNA, the US Naval Academy (Annapolis, MD) for ChiQat-Tutor-v1. These evaluations comprise 219 students for ChiQat-Tutor-v1 and 313 for ChiQat-Tutor-v2.

The third part, *Extending ChiQat-Tutor*, demonstrates the flexibility of ChiQat-Tutor, which can easily be extended to many other topics. Chapter 6 describes the existing extensions we ourselves developed: one additional data structure, Binary Search Trees (BSTs) and one computational process (recursion). Whereas the BST module was only piloted with a small number of students, the recursion version of ChiQat-Tutor was evaluated with more than 150 students, at two institutions: UIC and Carnegie Mellon University, Qatar campus (CMUQ). Chapter 7 provides a practical guide to implement a new module in ChiQat-Tutor from scratch, further demonstrating the flexibility of our approach and of the implemented architecture. Finally, in Chapter 8, we summarize the content of the book and further speculate on where the future can take ChiQat-Tutor.

Of note is that ChiQat-Tutor is available at http://www.digitaltutor.net/ and has about 6000 registered users besides the roughly 530 who participated in our experiments. This is another way in which our work has affected introductory CS education.

CHAPTER 2

Related Work

With Stellan Ohlsson

University of Illinois at Chicago

CONTENTS

THE design of software systems to support instruction is an interdisciplinary enterprise. The work presented in this book draws on four distinct disciplines: cognitive theories of learning; linguistic theories of dialogue, specifically, pragmatics; the pedagogy of Computer Science and educational technology, specifically, intelligent tutoring systems (ITSs).

The first disciplinary pillar of our work is the science of learning. Cognitive psychologists have long believed that pedagogical strategies and procedures can be made more effective by grounding them in a theory of learning. Prior to the emergence of cognitive science in the second half of the 20th century, this belief received little or no support from psychological research, presumably because the learning theories from which the instructional implications were derived were inaccurate. Neither Skinnerian rules of reinforcement [119] nor Piagetian stage transitions [337] lead to effective pedagogical techniques. Indeed, there is as yet no universal agreement regarding the basic cognitive processes involved in the acquisition of knowledge and cognitive skills. In the work reported in this book, we adopted a working hypothesis that we refer to as the *Multiple Modes Theory* [328]. The key idea of this theory is that learning is a cumulative result of multiple processes, including reasoning by analogy and studying worked examples, but also subgoaling and declarative-to-procedural transformations. We refer to these as modes of learning. There is no agreement as to which modes contribute most to the acquisition of cognitive skills. We assume that the modes are not mutually exclusive, and that different modes of learning are more or less useful in different circumstances [139, 327].

The second disciplinary basis for our work is focused on the communication between the student and the instructor. Our premise is that individual cognition is shaped by social interactions [101, 294, 337, 422, 353, 204]. The main methodological implication of this stance is the need to record and analyze linguistic interactions between instructors (tutors) and students. Linguistics provides theoretical tools and methods for analyzing utterances, including exhortations, explanations, questions, feedback messages, praise, criticisms, statements of goals and subgoals and other pedagogical uses of language. This type of analysis is grounded in pragmatics, the linguistic subdiscipline that studies how context contributes to meaning. Of particular relevance to our work is speech act theory [26, 378, 379], and its influence of analysis of tutorial interactions [133, 112, 170, 123].

Our third and fourth pillars both reside with Computer Science, in two different guises. As a third pillar, there is the subject matter to be taught and mastered, and its pedagogy [280, 343, 142]. Computer scientists have identified the key concepts that form the basis of computation and programming, such as computational thinking, data structures and algorithms. Our work asks how these fundamental concepts can best be taught, as informed by current insight into the pedagogy of teaching these concepts to beginner students at the college level. As the fourth pillar, Computer Science also provides the tools with which we can specify and build educational systems [415, 10]. For example, many instructional computer systems, especially Intelligent Tutoring Systems, were based on rule-based architectures, a well understood type of Artificial Intelligence system, and more recently, have adopted data mining and machine learning as a way to infer patterns from student behaviors, including within tutorial interactions. Hence, Computer Science plays two distinct roles in our enterprise: as a specification of the target subject matter, and as a tool kit for building the desired system.

In summary, our work is grounded in four distinct scientific disciplines. The remaining sections of the present chapter review in more detail the concepts and findings from these four disciplines that have played a central role in our work.

2.1 COGNITION AND MULTIPLE MODES OF LEARNING

In this section, we discuss nine modes of learning that have been studied in cognitive psychology research.

2.1.1 Background

The scientific study of learning began with the works by Thorndike and Pavlov toward the end of the 19th century [328]. The theoretical vocabulary available at the time was too impoverished to allow them to go beyond common sense verities such as the need for practice. Little progress was made in the first half of the 20th century. The cognitive revolution in the late 1950's expanded researchers' theoretical arsenal to include symbolic knowledge representations, the declarative-procedural distinction, hierarchically structured task strategies, goal-subgoal trees, feedback, heuristic search, as well as other concepts. But, learning received little attention in the first three decades of Cognitive Science.

This situation changed radically in 1979 when Anzai and Simon published an article reporting the first computational model of human skill acquisition through repeated practice on the so-called Tower of Hanoi puzzle [22]. Their model took the form of a computer program, written in Lisp, that could construct for itself a correct strategy in the course of solving the puzzle four times. The hypothesis behind the model was that during practice, the learner would move from "obvious" strategies, to more efficient ones; the cognitive question was how one strategy could be transformed into another. Anzai and Simon first engaged one single human subject who solved the puzzle repeatedly, and was instructed to think aloud. The authors then constructed a program that in a first incarnation, could simulate the four strategies the subject had adopted; but that further, in a second incarnation, could infer new strategies similarly to what the human subject had done. Specifically, as they describe it [22, p.132], they devised *an adaptive production system that uses similar processes to progress from an initial primitive search strategy to a strategy that avoids repetition of moves, a means-ends strategy, and an inner-directed recursive [recursive sub-goal] strategy*. They note that even if the system did not actually follow the temporal course of the subject's learning, it used the same type of information as the subject used and inferred three of the strategies that the subject had arrived at.

The particular hypotheses about skill acquisition implemented in Anzai and Simon's model were less important than the paradigmatic approach: collect verbal protocols from human subjects; hypothesize-specific computational learning mechanisms; implement them within an A. I. system; run the system repeatedly on the target task and compare the succession of changes in the model's behavior across training trials to the succession of changes in human behavior on the same task. If the model of the learning processes in the human brain is correct, then the predicted changes should mimic or approximate the observed changes in human behavior.

The Anzai and Simon's paper revolutionized the scientific study of skill acquisition through practice. Symbolic programming languages (Lisp, Prolog, etc.) constituted a formalism for the precise statement of hypotheses about the learning mechanisms that enable a learner to improve during practice. Simulation (i.e., the execution of a model) provides a way of deriving the behavioral implications of such hypotheses in an objective manner. The analyses of verbal ('think-aloud') protocols completed the new research paradigm by supplying a method for fine-grained testing of such implications against empirical data. The result was an unprecedented explosion of theoretical work on the acquisition of cognitive skills through practice. In the 40 years since the publication of their paper, more novel hypotheses about skill acquisition have been proposed than in the previous history of psychology [327].

A key question is how many distinct modes of learning a theory of skill acquisition should specify. In the immediate aftermath of Anzai and Simon's 1979 paper, strong claims were made to the effect that this or that mechanism was sufficient to explain all empirical phenomena associated with the acquisition of cognitive skills. Chunking of elementary pieces of skill knowledge into larger components is one example [365]. Constraining the applications of skill components (rules) in response to self-detected errors [27, 325] and reasoning by analogy [149, 126, 151] were alternative proposals.

Empirical tests that at first seemed straightforward turned out to have less power to discriminate among hypothesized learning mechanisms than expected. For example, the negatively accelerated learning curve observed in empirical data – the so-called power law of learning – was at first believed to be the signature of a particular type of learning mechanism, but it turned out that power law speedup of task performance is a characteristic of many types of change mechanisms [316, 18, 98, 355, 191]. As a consequence, every researcher could find empirical justification for his or her favorite hypothesis, and scientific progress through empirical testing began to seem less certain. The invention of novel learning mechanisms began to slow down.

In conclusion, there is no direct empirical support for the idea that all changes in behavior that occur in the course of skill practice are consequences of a single learning mechanism, nor is there any compelling theoretical reason to believe that all empirical phenomena associated with skill acquisition can be explained by one and the same learning mechanism. The alternative hypothesis is that multiple modes of learning are at play, to which we turn now.

2.1.2   Nine Modes of Learning

The alternative to a simple theory that derives all observable changes in behavior from a single process is a theory that claims that there are multiple distinct modes of change. We refer to this working hypothesis as the *Multiple Mode of Learning* principle. It raises the question of how many distinct modes of learning are there; how a learner benefits from each and how each mode can be supported by instruction.

Analysis of published computational models of skill acquisition revealed nine distinct ways that a task strategy can change and improve in the course of repeated training trials [327]. We summarize them briefly. They are ordered according to the three phases that have been broadly adopted in cognitive science [327, 228, 405]: first, initial learning that can apply before the learner has solved the target task even once; second, the feedback or fine tuning phase that relies primarily on feedback generated by and during practice and third, the automaticity phase that primarily produces speedup. We will briefly describe them here and then further elaborate on those that we focused on in our work.

Initial Learning Phase

1. **Advice taking.** In this mode, the learner receives direct instruction from an instructor or tutor about what to do (*when the green light comes on, press the red button*). The learner's task is to operationalize the advice/instruction; that is, to translate the tutor's discourse into executable mental code [15].

2. **Reasoning from declarative knowledge.** In this mode, the learner is in possession of factual knowledge that is not sufficient by itself to generate correct task solutions but nevertheless enables the learner to deduce the correct action in some situations [322].

3. **Goal analysis.** Skilled activity is purposeful; it aims to accomplish some goal or achieve some desired effect or outcome. The goal can be self-generated, posted by a coach or tutor or inhere in the definition of the task. (By definition, a person is not playing chess unless he or she is trying to check mate his or her opponent.) Regardless of origin, the structure and content of the goal is a source of information about how it is to be achieved. For example, the goal to make coffee contains the information that coffee is involved, and the goal to drive to location XYZ suggests that a car will be needed, implying that to search for coffee beans or locate a vehicle are plausible next moves. The process of decomposing a goal into one or more goals of lesser scope - subgoals - is often called subgoaling [299, 141].

4. **Worked out examples/demonstrations.** In this mode, the learner has an opportunity to study the solution to an example of the type of

task performance that he or she aims to master. His or her task is to translate the specific steps in the solved example into skill elements that are general enough to apply to future problems of the same type [413].

5. **Analogy/transfer.** In this mode, the learner has previously mastered a skill that is analogous to the target skill. The learner's task is to retrieve the analogue from memory, map the analogue to the current task and infer how the current task is to be dealt with. In some cases, the previously mastered skill and the current target skill are so similar that skill components are shared between them, in which case no work is required by the learner other than to notice the similarity and retrieve the previously mastered skill (identical element transfer) [149, 421, 126, 371, 151].

Feedback Phase

6. **Positive feedback.** As this term is used here, positive feedback consists of information that the learner is on the right track; his or her actions are appropriate, correct or useful. Such information can arrive either as a causal consequence of the learner's action (*the green light came on*), or as a verbal message from an instructor or tutor (*that was a good move* [302]). We hypothesize that to learn from this type of information, the learner must execute at least the following processes: (a) recognize the positive feedback as such; (b) create a knowledge structure that recommends the action performed under the relevant conditions or alternatively, (c) increase the priority associated with the relevant knowledge structure to make its recommendation less tentative.

7. **Error-correction (negative feedback).** In this mode, the learner perceives that he or she has made an error. The learner's task is to constrain the relevant skill elements so that he or she only executes the skill elements in situations in which they do not cause errors. This mode of learning makes use of error signals: environmental events that tell the learner that a tentative problem solving step was inappropriate, incorrect or unproductive. We refer to such error signals broadly as *negative feedback*, to include the interlocutor in the learner's environment. We speculate that the following cognitive processes drive this mode of learning: (a) recognizing error signals as such by the application of constraints on correct or good solutions and (b) restricting the application of the error producing action to a more narrow set of situations in which the action does not generate negative outcomes. Specifically when procedural skills are represented as rules, such "restriction" can come in different guises: from decreasing the strength of the corresponding rule [212]; to learning *critics* (or *censors*), rules that vote against performing an action during conflict resolution [40]; from extracting new rule conditions from the mismatch between constraints and the outcomes of

actions [325, 324]; to discrimination, namely, comparing successful and unsuccessful rule applications and adding discriminating features to rule [255].

Automatic Phase

8. **Short-cut detection.** In this learning mode, the learner makes use of memories of past solutions stored in memory. By comparing past solutions with each other, the learner can gather information about regularities in the problem space. These can be exploited to eliminate unnecessary steps from solutions, making problem solving more efficient. A famous short-cut detection in children's arithmetic is the so-called SUM-to-MIN transformation, in which children discover that it is unnecessary to count out the larger addend when adding small numbers [213].

9. **Quantitative regularities.** There is good evidence that the human brain is wired to recognize quantitative regularities in the environment. Examples include the relative frequencies of words, and the probability of particular action outcomes in certain situations.

2.1.2.1 *Discussion*

There might be other modes of learning not represented in this list. However, it seems unlikely that the list can be shortened: common sense as well as empirical observations agree that people learn in at least these nine ways. The different modes of learning differ with respect to what type of information they require as input. They also differ with respect to the complexity of the learning mechanism triggered. Advice taking often requires very complex translations from a declarative to procedural knowledge representations, while identical elements transfer anchors the other end of the complexity scale.

For our purpose, a key question is how learning modes reveal themselves in human-to-human interaction data. Evidence for these modes includes the environment signals that may trigger them, and the interactive tutorial strategies that can support them. Instructional and dialogue strategies are part of those *overt activities* that are in turn indicative of *potential* cognitive processes.

A second key question from an educational technology point of view is which modes should be included in such technology, and specifically, in Intelligent Tutoring Systems. One way to increase the pedagogical power of a tutoring system would be to provide support for all nine modes of learning, or at least, the seven that pertain to the first two phases of learning – as far as we know, the automatic phase has not been included in research focused on interactive learning activities, since it offers few if any options for supporting the learner's from that point of view; when a learner reaches the automatic phase, it is only a matter of extensive sequences of practice trials.

To the best of our knowledge, no system that supports all seven modes of learning in the first two phases has been implemented as yet. In the work reported in this book, we choose to advance tutoring technology according to the following working hypothesis: all modes of learning are more powerful when supported by human-to-human dialogue. This stance raises the question of what the effective aspects of such dialogues are, and how they can be implemented in a tutoring system or other type of educational technology. In the next section, we review the salient concepts and principles from research on linguistics and the study of dialogues that we put to work in the design of ChiQat-Tutor.

Before we turn to that section though, we will spend some more time discussing four of the modes of learning we listed above: positive and negative feedback, worked-out examples and analogy. These four will figure prominently in our tutorial dialogue analysis and in the features we incorporated in ChiQat-Tutor. Please note that the order we discuss them below is different from the order we presented them in above, which followed phases of learning. The order here is related to the complexity of their potential linguistic correlates, which can be even just a single word for *positive/negative feedback* but must be at least one sentence if not more for *worked-out examples* and *analogy*.

Positive feedback. As we noted earlier, positive feedback consists of information that the learner is on the right track, information which indicates that a problem solving step was appropriate, correct or useful [34, 102, 302]. Positive feedback is inconsistently named in the literature: strengthening, bottom-up knowledge generation or compiling results of search [315, 414, 396]. A general observation is that the effect of this mode of learning is uncertainty reduction, in that the learner may have performed a problem solving step because it appeared to be correct, or even choosing randomly, and positive feedback tells them they are in fact on the right path. Although verbal feedback is called "instruction" in the everyday use of that term, it differs from direct instruction as earlier defined in Mode 1. In the case of direct instruction, the learner has not yet performed the relevant action and is given help in deciding which action to take. The information arrives first, the action follows. In the case of positive feedback, the learner decides which action to perform, perhaps by guessing, and then performs the action. The new information resides in its consequences. The action comes first, the information follows. This difference impacts the relevant cognitive processes, so learning from positive feedback is distinct from learning from direct instruction, even though both the types of discourse are commonly called "instruction."

Error correction/Negative feedback. As we noted earlier, learning from errors [325, 324, 328] makes use of error signals: environmental events that tell the learner that a tentative problem solving step was inappropriate, incorrect or unproductive. We often refer to such error signals broadly as *negative*

For the equation $a = ag + b$, express a in terms of the other variables.

$a = ag + b$

$a - ag = b$

$a(1 - g) = b$

$a = \frac{b}{1-g}$

Figure 2.1 Worked-out example [400].

feedback, to include those provided by the interlocutor in the learner's environment, for example, the corrections and criticisms that a human tutor provides during tutoring. We should note that negative feedback has other effects on the learner beyond affecting the development of skills: according to social cognitive theory, negative feedback is harder to accept than positive feedback [92] because it is self-threatening, and especially those who could make the most use of it, ignore or resist it [25]. At the same time, through negative feedback, people can take corrective action that contributes to their personal growth well beyond learning a skill [220]. Even ignoring social determinants, research in neuroscience shows that is it is only starting around 11 years of age that humans start to be able to make use of negative feedback in an effective way [412]: *the neural activation patterns found in 11- to 13-year-olds indicate a transition around this age toward an increased influence of negative feedback on performance adjustment.*

Worked-out Examples   The term *Worked-out Example* (WOE for short) was first coined by Sweller and Cooper [400]. WOEs, also called 'worked' or 'solved' examples, provide a step-by-step example in solving a problem, as shown in Figure 2.1. Sweller and Cooper found WOEs to be an effective teaching strategy, even preferable to regular problem solving for novices.

Further, Sweller investigated the reasons for the effectiveness of WOEs [399]. The fundamental assumption is that the foundations of domain specific knowledge lay in the form of schemas. These schemas are at the core of problem solving, and can be described as a cognitive structure, that allows problem solvers to recognize the type and state of a problem. From this state, and understanding the goal state, it is therefore possible to arrive at a solution via a number of steps. If a human does not hold a schema for a particular problem, it tends to be very difficult to identify the type of problem or make rational decisions on how to arrive at the desired solution.

Based on this, Sweller claims that conventional problem solving does not give the best opportunity for learners to acquire these schemas. Therefore, problem solving may not be a wise choice for someone to learn a new topic. The reasons for why schema acquisition is not optimal in problem solving are based on cognitive load theory. This theory is centered on the idea of working memory in humans [298, 299], namely, that part of human cognition is devoted as a temporary store, much like a scratch pad. This store can receive information, manipulate it and works with other cognitive subsystem, such as short-term memory. Importantly, working memory is where humans process complex information and therefore is needed for learning. Unfortunately, working memory is not an infinite resource and differs between individuals: trying to perform too many activities at once may exhaust working memory and therefore result in cognitive overload, which may hinder learning. Sweller claims that learning instruction should be tailored toward the availability of resources.

It is thought that worked-out examples may be more effective than problem solving alone due to their potential of reducing certain types of cognitive load. Problem solving naturally places a far higher load on the learner, since they have to understand new material, acquire new knowledge, commit it to memory and solve a specific problem at the same time. Bearing in mind that working memory may only be able to process two or three novel concepts at once, any reduction may aid in learning. Examples take away the additional extraneous overhead of problem solving and allow more cognitive resources to be allocated to germane cognition. This will allow the learner to build the appropriate schemas more efficiently. Once the schemas are built, the learner could reinforce this learning with additional problem solving, which no longer includes the cognitive overhead induced by the need to create these schemas.

Although examples appear to be a useful strategy in reducing cognitive load, there is the possibility of introducing *expertise reversal* [219]. As just discussed, examples may be more effective at creating schemas which is ideal for beginners. For more advanced students, examples may not be beneficial since they already have acquired such base knowledge and problem solving may be a more viable alternative. Kalyuga et al. [219] noticed that advanced students may also be hurt by being given examples. This phenomenon suggests that not all student should be given examples and that students' knowledge should be evaluated before example deployment to make sure they learn efficiently. In Section 5.3.1, we will return to the issues of when and for which students WOEs may be more effective.

Analogy above, we listed analogy as one of the modes of learnings: for some researchers, analogy is veritably the core of cognition [198]. The cognitive theory of analogy has been developed in the last 35 years mostly by researchers like Holyoak [156, 157, 199, 200, 53] and Gentner [149, 126, 151, 150, 152]. Among the different theories, we briefly review Gentner's, because of its well-defined impact on computational models (in the sense of Computer Science, not only of Cognitive Science). General reviews of research on analogy in Cognitive Science can be found in [207, 233].

According to Gentner [150], *analogies are partial similarities between different situations that support further inferences. Specifically, analogy is a kind of similarity in which the same system of relations holds across different objects. Analogies thus capture parallels across different situations.* In her view, a theory of analogy must describe how the meaning of an analogy is derived from the meanings of its parts, which she embodies in *Structured Mapping (SM)*. SM includes rules for mapping knowledge about a base domain into a target domain. Importantly, these rules do not depend on domain content but only on syntactic properties of the knowledge representation. The mapping rules map relations between objects from base to target domain, rather than mapping attributes of objects. For example, given the analogy "An electric circuit is like a plumbing system" [154], their relational similarity is based on a common causal structure; in both the cases, current/water flow is higher if voltage/pressure is higher. Note that the analogy holds no matter the kind of pipe the water flows through/the kind of wire the circuit is made of.

Structural mapping consists of three steps. First is retrieval: given some current topic in working memory, a person may remember or be made aware of an analogous domain which they already possess (or are presumed to possess) in long-term memory. Once both the cases are present in working memory, mapping, the second step, can happen: during mapping, representations are aligned and inferences are projected from one analog to the other. The third step is evaluation: once an analogical mapping has been done, the person will judge the "quality" of the mapping, in terms of structural criteria, validity and relevance.

The structural mapping approach has several computational incarnations (again, with computational we refer to artificial computational processes). [116] proposed the Structure Mapping Engine (SME):

> SME constructs all consistent ways to interpret a potential analogy and does so without backtracking. It is both flexible and efficient. Beyond this, SME provides a "tool kit" for building matchers that satisfy the structural consistency constraint of Gentner's theory.

SME is used as a simulation engine in the sense we described earlier in Section 2.1.1, to both verify and fine-tune a cognitive theory, and keeps being modified and updated [125]. Other computational models of Gentner's theory of analogy can be found in the connectionist tradition [153], the precursor of today neural networks and deep learning.

Like for Worked-out Examples, we will come back to analogies in the context of ChiQat-Tutor in later chapters.

2.2 PRAGMATICS AND DIALOGUE PROCESSING

In our work, we analyze how pedagogical strategies that are grounded in the multiple mode of learning theory are expressed in one-on-one dialogues between a tutor and a student. At the same time, these dialogues obey most of the same rules as ordinary human conversation. Hence, linguistic theories of conversation, and especially pragmatics, have also influenced both our analysis of the human data and our computational modeling.

Pragmatics is the branch of linguistics that studies the usage of language in context, or informally, "language beyond the single sentence". As concerns human conversations, insights from philosophy of language going back 50 or 60 years provide the foundation of the theory of speech acts, which capture the speaker's intent behind an utterance. The idea that utterances in a conversation are actions performed by the speaker was initially due to [26, 378, 379]. These philosophers distinguished between direct speech acts, i.e., the literal interpretation of the utterance, and indirect speech acts, i.e., the true intent of the speaker. To wit, compare the two utterances: *can you parachute?* and *can you mail this letter for me?*. Literally, they are both questions; however, the second exhibits an *indirect force*, namely, the true speaker's intention is to give a command, couched as a polite request.

Various sets of speech acts have been defined over the years, and speech acts, especially indirect speech acts, are at the basis of a number of early conversational systems [79, 78, 11, 424]. In the last 20 years or so, the idea of speech act has been operationalized from a computational perspective, as a dialog act (DA). Beyond the difference in nomenclature, the notion of a dialogue act also accounts for other influences on the theory that includes the ideas of adjacency pairs, pre-sequences and other aspects of the intentional properties of human conversation developed in the field of conversation analysis ([260], [217, ch.24]).

This operationalization resulted in many plausible inventories of DAs, including an ISO standard [58, 409, 385, 19, 51, 52]. Such inventories have in turn been the foundation for the development of dialogue corpora, annotated with one of those proposed inventories. The annotated corpora have been used to develop computational models of dialogue first based on sequential models such as Hidden Markov Models (HMM) [394], Conditional Random Fields (CRF) [403, 64] or Reinforcement Learning (RL) [387, 118] and more recently, based on neural networks/deep learning models, in some cases combined with the more traditional HMM, CRF or RL [426, 250, 5].

Closer to the topic of this book, conversation analysis in general and dialogue acts, in particular, have been used to model tutorial dialogue as well [133, 170, 259, 364, 111, 266, 45, 429, 368, 273], including by us, as we will discuss in Chapter 3. Before we move to discussing this body of literature, we will briefly hark back to cognitive theories of learning, by noticing that some cognitive and education researchers stress the primacy of social interaction on individual cognition [101, 294, 337, 422, 353, 147, 204], where of course much social interaction is mediated via language. This is not incompatible with but rather orthogonal to the Multiple Modes of Learning we discussed earlier, in that as we noted earlier, a key question for us is how learning modes reveal themselves in human-to-human interaction, and specifically, *one-on-one tutoring*. The importance of the latter was established by the famous 2σ effect paper by Bloom [41], who found that the performance of one-on-one tutored students, as opposed to the performance of students taught in the classroom, increased by about two standard deviations, a very large effect. Also, relevant to the analysis of tutoring conversations is Chi's framework [66, 69] to compare *active, constructive* and *interactive* learning activities, even if crucially in Chi's approach, *interactive* is not to be taken in its usual meaning, of *mutually or reciprocally active* (between people), or *involving the actions or input of a user* (between a user and a computer) (definitions from the online Merriam-Webster dictionary). Namely, *interactive* here refers to learning activities, not to the style of engagement. For example, during a tutoring dialogue, a student can be *active*, doing a physical activity, such as highlighting a passage, or gazing at a window on a screen; *constructive*, such as explaining or elaborating or *interactive*, namely, engage in guided construction by responding to scaffolding. Chi then appealed to various empirical studies to argue that interactive learning activities are more likely to support learning than active or constructive activities. Notwithstanding the meaning of *interactive* in Chi's systematization, it is still the case that face-to-face conversations are the most frequent vehicle for interactive learning activities, hence closing the circle back to one-on-one tutoring, and tutoring dialogues, to whose analysis we now move back.

With respect to the more "conventional" dialogue analysis literature within linguistics and computational linguistics, the modeling of tutoring dialogues has brought at least three research issues to the forefront: which dialogue acts appear in tutoring dialogues and with which meaning; which sequences of dialogue acts constitute tutoring strategies and how dialogue acts, either individually or as part of tutoring strategies, correlate with learning.

The first issue arises from the specific *genre* of tutoring, in which some of the regular conventions of dialogue may be flouted: a classic example is that tutors ask questions although (presumably) they already know the answer. Additionally, detailed inventories of DAs in tutoring are developed precisely to find correlations between fine-grained behaviors and learning outcomes. Hence, various studies have highlighted which specific types, or subtypes, of DAs occur in these dialogues, for example, what type of DAs student use (e.g., *request for confirmation, request for information, challenge, refusal to answer,* and *conversational repair* [382, 111]); or which type tutors use (for example, among the broad notion of tutor statements, *restatement, recap, request, bottom-out, hint, expansion* [266]).

The second issue pertains to whether sequences of dialogue acts result in higher-level pedagogical strategies. For example, [45] derives an HMM with 8 states for one tutor and 10 states for a second tutor; each state in the HMM represents a specific *tutoring mode* that both responds to the student's problem solving state, and has a certain probability to be reached from another mode. [273] derives a Markov process for the entire session, composed of states (also called *modes*) such as *Fading, Scaffolding* and *Rapport Building* - each mode in turn may be composed of several dialogue acts, for example *a sequence of hints in the form of questions may reflect a scaffolding instructional strategy.*

Finally, the overriding question is always which dialogue acts or dialogue acts sequences may result in increased learning. Almost all the references mentioned above address this question. For example, [266] presents a detailed investigation of correlations between pairs of tutor/student moves and student learning. Among their findings, a student novel answer followed by a tutor bottom out is highly positively correlated with learning, whereas a tutor hint followed by a student shallow answer is highly negatively correlated with learning. In the models discussed in the previous paragraph, some of the HMM states derived by [45] as a model of the dialogue are then found to be correlated with learning; on the other hand, in [273], the modes are discovered precisely because they (or their subcomponents) highly correlate with learning.

As we will discuss in Chapter 3, we will also address all three issues: which DAs appear in tutoring dialogues; whether sequences of DAs comprise higher-level tutoring strategies and the correlation of DAs and higher-level tutoring strategies with learning outcomes. However, one difference between most of the literature discussed so far and our work is that the DAs we will annotate for are grounded in a specific cognitive theory of learning, the Multiple Modes we discussed above. Additionally, rather than deriving pedagogical strategies bottom up as in [45, 273], we will investigate the higher-level strategies motivated by the Multiple Modes of learning theory; only then, will we analyze the DAs they are composed of as a way of operationalizing those higher-level strategies in ChiQat-Tutor, our ITS.

2.3 INTRODUCTORY COMPUTER SCIENCE EDUCATION

The demand for Computer Science mastery is at an all-time high, with the need for skilled practitioners far exceeding availability – the US department of Labor projects the demand for software developers to increase 26% from 2018 to 2028.[1] In fact, in the last few years, the number of Computer Science (CS) majors has skyrocketed [444], reversing the dramatic downward trend from the early 2000s. The increase in enrollment at the undergraduate level is such that even the National Academies of Sciences, Engineering and Medicine issued a report in 2018 with recommendations for best practices [313].

Beyond CS majors per se at the college level, foundations in computing are now seen as necessary for everybody to join the work force, and more in general, for everybody to be an informed citizen of our increasingly technological world. Among the first proposals to advocate for foundations in CS to be part of the knowledge of an educated citizenry was *Computational Thinking* in the early 2000s [432, 433, 99]. As initially stated in [432], computational thinking is *a way of solving problems, designing systems and understanding human behavior that draws on concepts fundamental to computer science;*

---

[1] https://www.bls.gov/emp/tables/fastest-growing-occupations.htm. Accessed June 2020.

computational thinking is grounded in the human ability to choose abstractions, operate with them and automate them via algorithms. Whereas the idea of computational thinking goes back to many earlier thinkers such as Papert's 1980 book [332], the proposals in the early 2000s coalesced various strands of thinking around how to best introduce CS foundations in K-12 (the elementary plus secondary education system in the USA, from Kindergarten to grade 12). The ensuing further debate of these concepts [314] touched on many aspects, including whether to provide full in-school curricula, and not only extra-curricular activities [376]; it contributed to an intellectual climate that today takes it for granted (or almost so), that CS foundations should be included at least in the high school curriculum. Educational policies are now designed to that effect, including at the highest level: the initiative *Computer Science for all* was launched by President Obama's White House in 2016.[2]

In the following, we will look at initiatives for introducing CS in K-12, including recent pedagogical trends that attempt to really turn CS into a subject *for all*. We will then switch to recent trends in college level CS education.

2.3.1 Elementary and Secondary Education

As noted earlier, computer science and information technology have penetrated all aspects of modern society, and therefore, there is a push for Computer Science for all, in that every citizen should have enough insight into technology so as to make informed choices. As a consequence, various initiatives on introducing CS to school age students (K-12) have been proposed in the USA, with a push to move the introduction of such concepts even earlier, in pre-K. Initiatives run the gamut from National Science Foundation's call for proposals under the *Computer Science for All* theme, to fully developed curricula that include teacher development initiatives such as *Exploring Computer Science (ECS)*[3] and *Computer Science Principles*.[4] Similar initiatives are taking place in the European Union to increase the digital prowess of its population: for example, the Digital Skills and Jobs Coalition[5] launched in 2016 is *a multi-stakeholder partnership to bolster computer science proficiency at all levels within Europe's workforce pipeline*. While this European Union initiative is not focused on K-12 or on broadening participation specifically, some of the efforts focus on bridging the digital divide for women and girls, or for refugees. As concerns AustralAsia, we have not been able to find initiatives encompassing more than one country; however, for example, the Hour of Code that we will illustrate below is very popular in countries such as China, Vietnam, Malaysia and Singapore.

---

[2]https://obamawhitehouse.archives.gov/blog/2016/01/30/computer-science-all. Accessed June 2020.

[3]http://www.exploringcs.org/

[4]https://code.org/educate/csp

[5]https://ec.europa.eu/digital-single-market/en/digital-skills-jobs-coalition

Since the mid 2010s, in the USA, there has been a keen and increasing interest in bringing CS to elementary school children [343]. The motivations are twofold: first, considerations of equity and accessibility, that we will further discuss shortly as concerns the ECS curriculum in high school, and second, the value of Computer Science per se as a building block, since *Computer science has [far more] to do with problem solving, persistence, and logic [than it has to do with keyboarding]*, and *a strong and lasting foundation for computer science must include the development of critical thinking and decision-making skills, as well as self-confidence* [343, pp. 61–62]. Among various initiatives to reach elementary and middle school students, Code.org is a nonprofit devoted to make CS more accessible in schools, especially to girls and underrepresented minorities (URM). *[Their] vision is that every student in every school has the opportunity to learn computer science, just like biology, chemistry or algebra.*[6] Code.org has developed several curricula, starting with pre-readers (age 4) all the way up to age 13.[7] At the high school level, Code.org is a *provider of curriculum and professional development for AP Computer Science Principles*, where *AP* stands for *Advanced Placement* and refers to courses that high school students can use to place out of introductory courses in the specific discipline in college. Code.org is also a partner with the *Hour of Code* events. The Hour of Code started in 2013 as a one-hour introduction to computer science, *designed to demystify "code", to show that anybody can learn the basics, and to broaden participation in the field of computer science*. It is now a worldwide endeavor to promote CS, which still starts with one hour coding activities but then expands to all types of community efforts. Over 100 million students all over the world have participated in one of these events.[8] Many papers, both peer-reviewed, and in the popular press have addressed the effectiveness of both the Code.org curricula [251, 439, 360] and the Hour of Code efforts [408, 106, 274], in many countries in the world (from Serbia to Taiwan, from Hong Kong to of course the USA). Whereas our book is not the venue to review these results, informally one can conclude that both formal curricula adoption and participation in the Hour of Code have exponentially increased in the last 5 years, and have reached more under-represented groups (girls and minorities). The jury is still out on the long-term effects of these initiatives on future college enrollment and the workforce, especially for underrepresented populations. One very promising statistics is that in 2017, the first year in which this information was available, almost 12,000 Code.org students took and passed the AP (Advanced Placement) CS Principles exam, of which one-fourth female and about 20% URMs; in 2018, the total number raised to about 19,500, with the percentages of females rising to 33%, whereas the percentage of URM remained more or less constant.[9]

---

[6]https://code.org/about. Accessed June 2020.
[7]https://code.org/student/elementary. Accessed June 2020.
[8]https://hourofcode.com/us#faq. Accessed June 2020.
[9]https://code.org/statistics. Accessed June 2020.

We will now illustrate in some more detail efforts at the high school level via ECS: this curriculum is offered among others in the Chicago Public Schools (CPS), to which UIC is linked via numerous collaborations and initiatives, both as concerns supporting education in CPS and recruiting CPS students as freshmen at UIC. ECS is a member of the national CSforAll consortium,[10] which aims to increase access to and participation in CS across the USA.

In the late '90s and early 2000s, ECS was spurred by a keen interest in understanding why so few students of color and women would study CS either in high school or in college [120, 279, 278], which prompted the researchers (initially led by Goode and Margolis), to ask questions about more accessible and inclusive curricula [161, 280], and about how to better involve teachers in curricula development [162, 160, 163]. From the beginning, this curriculum [280, p.73] *was purposefully designed for broadening participation in response to the finding of the first edition of* [278]. To do so, the ECS curriculum covers the breadth of CS through *inquiry-based, hands-on, culturally relevant instruction*; and crucially, it does so by providing a rigorous and supportive teacher professional development program, which includes: *training in CS content, pedagogy and belief system (including stereotypes about which students can excel in CS)*, and *participation in a teacher community for reflection, discussion, collaboration and support*. To put their money where their mouth was, Goode, Margolis and collaborators did not start by focusing on schools which attract mostly privileged students; rather, they partnered with LAUSD, the Los Angeles Unified School District, since it includes [280, p.74] *schools serving students who are underrepresented in the field of CS (in Los Angeles, this means predominately African American and Latino students)*. As of this writing, *the curriculum is currently taught in at least 34 states and Puerto Rico, including the 7 largest school districts, as well as some rural locations and reservations. Over 55,000 students participated in ECS courses nationwide in 2018-19, with an additional 6,000 expected in 2019-20.*[11]

The ECS curriculum comprises six units, ordered as follows: 1. Human Computer Interaction, 2. Problem Solving, 3. Web Design, 4. Programming, 5. Computing and Data Analysis and 6. Robotics (units 5 and/or 6 can be substituted by a unit on e-textiles or one on Artificial Intelligence.) These units focus on traditional foundational concepts such as algorithms, problem solving, data analysis, programming (with Scratch), but contextualize assignments and instruction so that they are socially relevant and meaningful for a body of diverse students. One fundamental tenet of ECS instruction is inquiry-based learning, a rigorous approach that calls on teachers to prepare students to ask critical questions; teachers then act as guides to help students steer their investigations. Another fundamental tenet is building on students *funds of knowledge and cultural wealth*. ECS places CS education in the context of students' communities, lived realities and interests, and each ECS unit gives

---

[10] https://www.csforall.org/
[11] http://www.exploringcs.org/for-teachers-districts/ecs-now, accessed in June 2020.

teachers opportunities to incorporate into CS, learning issues that are relevant to students' lives. For example, in one ECS class, students designed a web page about their grandparents' country of origin and personal stories; in another, students used Scratch to design video games that taught players a message that was personally meaningful to the student-designer, such as: how to promote healthy lifestyles, undocumented immigrants and their struggles, the high school to college pathway, cancer and more; in another project, students used their mobile phones to design surveys and collect data about food consumption in their neighborhood, trying to understand why it is considered a "food desert".

Many studies have been carried out about different aspects of the ECS curriculum efficacy. To start with, what are the predictors of failure in the ECS curriculum itself? One longitudinal study in CPS (Chicago Public Schools), one of the few school districts that has introduced one year of CS education as a requirement for graduation, found other predictors of failure in this course other than the obvious ones, low GPA and attendance [288, p.9]:

> Female students are less likely than male students to fail. Hispanic students are more likely to fail ECS relative to students who identify with other races. Correspondingly, students in an English language learner program (which in CPS is primarily comprised of Hispanic, Spanish-speaking students) were more likely to fail than native English speakers. Students in a special education program were also more likely to fail than students not participating in a special education program.

Additionally, this report found that not surprisingly, the more experience (in terms of repeating the course) a teacher had, the less likely were the students to fail; and that the probability of the students failing the class was higher if the teacher had not attended the rigorous professional development program associated with the ECS curriculum.

Another longitudinal study [289], again in the Chicago Public Schools, explored whether students completing the ECS courses were more likely to take other CS courses; however, this study took place from the 2011-12 through the 2015-16 school years, when ECS was still an elective and not required. After controlling for prior achievement and demographic variables, students who took an ECS class were more likely to pursue another CS course than students who started with a traditional CS course. For example, a female, Hispanic student with an average GPA who took ECS and received an average grade in the class had a 36% probability of taking another CS course in comparison to a person of the same ethnicity and gender, who took the traditional class, and who had only an 18% probability of taking another CS course.

The ECS project is truly remarkable in its conception, broad adoption across the USA and success in increasing the diversity of students studying CS in high school. It remains to be seen whether it has managed to achieve all its goals; this is probably an unfair question to ask, given how ambitious and transformative the ECS project is, and that it attempts to address systemic problems in access and equity that include many diverse stakeholders. Undoubtedly, many more studies and adjustments will shed further light on the effect on the ECS curriculum on students' attitudes toward CS, and on its effectiveness toward its important goal, that of broadening participation in computing.

2.3.2   From high school to college

Even with the current new approaches to introducing CS in high school, many students still arrive in college without a clear idea of CS as a field of study. Dan Garcia at Berkeley (who in the span of a distinguished 20 years career has received much recognition for his efforts in introductory CS education) has been a long time advocate of introductory courses for non majors, both in the spirit of *CS for all* and as a "gateway to the major". First offered in the late 2000s, the *Beauty and Joy of Computing (BJC)* course has since been adopted by Berkeley as the first CS course for non majors, and is offered every semester [290, 144, 389, 142]. BJC is also a CS Principles course in high school, where in fact much effort of the Garcia's group has focused in recent years, including on teachers' professional development [145, 137, 28]. Two ACM Inroads articles [389, 145] summarize the approach taken in BJC and its effects. Specifically,

> [BJC] is a CS Principles course whose guiding philosophy is to meet students where they are, but not to leave them there. It covers the big ideas and computational thinking practices required in the AP CSP [CS Principles] curriculum framework and powerful computer science ideas like recursion and higher-order functions. The programming part of the course uses Snap!, an easy-to-learn blocks-based programming language based on Scratch. [145, p.71]

BJC takes the approach that students must have fun: the course is based on guided programming laboratories where students are asked to explore and play. At the same time, BJC doesn't shy away from some of CS foundational ideas, such as algorithmic complexity, data structures (specifically lists), functions-as-data, recursion (via fractals) and higher order functions.

One fundamental tenet of BJC is that it is open to everybody, and in fact *One thing that defines CS10 [BJC] is the variety of people and stories coming together to experience computer science for the first time.*[12] BJC attracts a diverse crowd, especially as concerns gender: at UC Berkeley, it started enrolling more women than men in 2013, and in Spring 2018, it boasted its highest ratio of women to men (65% to 35%). Whether BJC will be effective as a gateway course still remains to be seen. [286] contrasted BJC (as CS10)

---

[12]https://cs10.org/su20/syllabus.html. Accessed July 2020.

and CS361A (*the Structure and Interpretation of Computer Programs*), which is a more traditional introductory programming and computer science course. The main methodology in [286] was to compare surveys taken at the beginning and at the end in each course for changes in sentiment, and usage of words associated with a stereotypical Computer Scientist (white or Asian male, nerd, etc). McConnaughey found that sentiment toward Computer Science became significantly more positive for CS10 students, while it became significantly more negative for CS361A students. The description of CS types via stereotypes became significantly less so for BJC students, while it stayed more or less constant for CS361A students. Further, gender did not affect changes in sentiment, but did affect changes in stereotypes: specifically, at the end, females in BJC had a less stereotypical view of CS, but unfortunately, the opposite happened to females in CS361A. Status as URM did not interact with changes in sentiment or in stereotypical views of CS. This study also found a correlation between positive sentiment and subsequent retention of female students (from BJC to CS361A, and from CS361A to CS361B).

As for the ECS curriculum we described in the previous section, the jury is still out to assess the impact of such courses on retention and on increasing participation of women and URM's in CS. Other initiatives to support underrepresented groups are still needed. For example, in [143], Garcia describes the CS scholar program to support women whose pathway into the CS major at Berkeley is via BJC and other two courses; these students need to obtain a 3.3/4.0 GPA in the introductory sequence. As Garcia notes,

> The performance of our scholars has been higher than the general population every year, and consistently rising. We still have room to grow, however, as the cohort's GPA average is not yet 3.3.

2.3.3 Post-Secondary Education for CS Majors

We now turn to more traditional, introductory CS education at the college level, as encoded in accredited curricula; some of the materials taught in such introductory classes are the target of ChiQat-Tutor.

The joint task force of the two preeminent CS professional associations, the ACM and the IEEE Computer Society, periodically revises the CS curriculum [1, 3].[13] Through various revisions, several fundamental concepts have persisted. One is algorithmic computation, where algorithms are seen as models of computational processes; algorithmic thinking is of course part of all the curricula we have discussed so far. Another fundamental concept is data models, i.e., standard information constructs for representing data, on which

---

[13]The latest CS curriculum revision is from 2013; a 2020 draft of the next revision is available at https://cc2020.nsparc.msstate.edu/. There are more recent official recommendations on areas such as cybersecurity (2017), Computer Engineering (2016) and Software Engineering (2014). https://www.acm.org/education/curricula-recommendations

algorithms operate. At a very abstract level, computers manipulate information, represented in a specific way. How data is best represented in computer memory depends on the processing that is to be done. In general, however, almost any computer application needs to *store, retrieve* and *process* information. Information storage representations are usually called *data structures*. The most fundamental data structures are linked lists (lists for short), stacks, queues and Binary Search Trees (BSTs) (they are all briefly introduced in Appendix A) – lists are part of the BJC course; the notion of trees is also introduced in some of the curricula we discussed earlier, via say genealogical trees, or minimal spanning trees for graph traversal, but not necessarily more specific types of trees and attendant algorithms, like BSTs.

Data structures often present difficulties for students, as many CS educators have noted [1, 2, 375, 3, 443]. Systematic studies of why difficulties arise and for which specific data structures are rare. Most research relating cognitive difficulty and CS either attempts to holistically inform the whole CS curriculum in the classroom [37, 264, 377, 136], or examines how novices approach programming [391, 434, 295, 352, 49, 50, 388]. At the same time, several studies provide indirect evidence for the common wisdom that students have trouble mastering static data structures that are manipulated via dynamic procedures [1, 404, 221, 443]. Additionally, as in physics or chemistry [235, 296], students must develop the ability to move seamlessly among multiple presentations of the same concept. For example, in textbooks, data structures are presented verbally, graphically and in a process-oriented way, via code or pseudo-code [208] (Figures 4.6, 4.7 and 4.8 in Chapter 4 show three representations of a list and its processing, graphical in the window on the bottom left, textual in the top left window, and the code in the window on the right). Understanding data structures inherently requires understanding the algorithms that manipulate them, and algorithmic thinking in general. Recursion is one inherently difficult algorithmic strategy to master [338, 340] and an additional source of confusion when algorithmic manipulation of data structures themselves can be iterative or recursive [285, 307]. Educational technology and, in particular, Intelligent Tutoring systems can help students in introductory CS courses precisely by highlighting the dynamic aspects of data structures, and by supporting students in understanding them, as we will discuss in Section 2.5.

To conclude, this overview of current trends in introductory CS pedagogy, we will observe that the difficulties we just discussed in understanding foundational CS concepts are one reason why undergraduate CS courses, whether for majors or nonmajors, often experience exceedingly high rates of attrition, especially at the freshman-sophomore transition, and especially for underrepresented groups [35, 222, 96, 117, 342, 39, 398, 347, 195]. Other reasons include whether the content is appropriately leveled, so that the students do not experience too much frustration [230]. Of course, other reasons pertain to the sense of not belonging that we discussed in the previous sections with respect to the ECS curriculum and the BJC course: when students from groups underrepresented in STEM choose to enroll in an introductory computer science course, they seldom find the topics engaging and relevant to their own lives [120, 398]. However, when faculty take students' interests and backgrounds into account, students' enrollment and persistence in CS courses increase [30]. A fourth-part series of short opinion pieces at the end of 2017 and beginning of 2018 by the ACM Retention Committee provide first an overview of several common problems and circumstances related to curriculum that may discourage students from persisting in computing majors [423]; then 12 tips for improving the culture in institutions, departments and classrooms that better support all students who attempt a computing major [262]; some suggestions on how to help individual students when they encounter situations that may require special intervention [354] and finally, suggestions on how to better engage URM students [420].

2.4 INTELLIGENT TUTORING SYSTEMS (ITSS)

ITSs have been around since the '80s, building on what was known as Computer-Aided Instruction (CAI), which had been around since the '70s, if not the '60s [56]. Research on ITSs is inherently interdisciplinary, being grounded in Computer Science, mainly in Artificial Intelligence and Human Computer Interaction; in cognitive psychology; and in educational research.

ITSs could be defined along multiple dimensions, but in general, they are considered to be

> computer programs that use AI techniques to provide intelligent tutors that know what they teach, whom they teach, and how to teach. AI helps simulate human tutors in order to produce intelligent tutors. [10]

This definition indirectly points out the difference between ITSs and CAI: in general, a CAI system will not monitor the learner's solution steps and give feedback [415, 10].

There is a broad agreement that ITSs must have at least three types of knowledge: about the content that will be taught, about the student and about pedagogical strategies; the latter may also include information about how to communicate with the students. Hence, a 'typical' ITS will include [10]:

> the domain model which stores domain knowledge, the student model which stores the current state of an individual student in order to choose a suitable new problem for the student, and the tutor model which stores pedagogical knowledge and makes decisions about when and how to intervene.

Often, the classic ITS architecture also includes a *user interface* module that allows the student to interact with the system. According to [392, p. ii],

> The user interface interprets the learner's contribution through

various input media (speech, typing, clicking) and produces output
in different media (text, diagrams, animations, agents).

A variety of approaches have been adopted for each of these modules,
which we will briefly describe, other than for the user interface, since the
latter is often idiosyncratic to the specific ITS. We will indirectly touch on the
user interface though when discussing the tutor model and how pedagogical
strategies affect the "signals" that the student can input and/or receive, such
as flags/graphics/NL interaction, whether typed or spoken.

The domain model (also called expert model) includes facts, rules,
problem-solving strategies in that particular domain. This domain knowledge is used both to present correct information to students and to evaluate
their performance. Two main approaches have been used to represent the domain model, cognitive tutors (model tracing) and constraint-based modeling
(CBM).

Cognitive tutors embody the cognitive model expounded by the ACT-
R theory of cognition and learning, initially developed by J. Anderson
[13, 14, 358]. ACT-R distinguishes between explicit (declarative) and implicit
(procedural) knowledge. Declarative knowledge consists of a network of linked
concepts, called chunks. Procedural knowledge is represented in the form of
IF-THEN production rules. In a cognitive tutor, a rule-based model generates
step-by-step solutions, where feedback can be provided to students on the correctness of each step; such model keeps track of many different approaches to
the final answer. Crucially, the domain model also includes common mistakes
that students make [17, 232, 84, 8, 231, 268].

A second popular approach is that of CBM, grounded in Ohlsson's theory of learning from performance errors [323, 326, 328]. Knowledge is represented through constraints that cannot be violated; as a consequence, correct
solutions cannot violate any domain principles. Constraints are represented
as IF-THEN rules: when the relevance condition (IF) holds, the satisfaction
condition (THEN) must be true. Instead of representing both correct and
incorrect knowledge as in model tracing, it is sufficient to capture only the
domain principles. CBM gave rise to its own suite of ITSs, including our own,
and those from the Mitrovic' group in New Zealand [301, 330, 300, 303, 65].

The student model is used to dynamically represent the knowledge and
skills of the student, either directly or through aspects of the student's behavior, in such a way the ITS can infer the state of the student's knowledge
and skills. There are a variety of approaches to represent the student model
[73], also because there are many purposes for which the student model is
used, including: corrective (to remove errors in students' knowledge); elabo-
rative (to fill gaps in the student's knowledge); predictive (to understand how
the student responds to the system's actions) and evaluative (to assess the
student's overall progress) [10].

A common approach to the student model is the *overlay model*, developed already in 1977 [59]. It simply assumes that student knowledge is a subset of domain knowledge; hence, any student's behavior different from what is stored is considered a gap. At least originally, the purpose of the overlay model was mainly corrective, namely, to eliminate the gap between those two models as much as possible. The simplest overlay models use binary logic; more recently, qualitative measures indicate the level of student knowledge (good, average or poor). With an overlay model, the ITS can aim to improve the student's knowledge as much as possible; however, neither alternative paths to solutions nor misconceptions play any role. Additionally, there is no room in an overlay model for students' individual characteristics and preferences. As a result, most current ITSs that use an overlay model combine it with other mechanisms [73].

Another common approach for student modeling uses Bayesian Networks, directed acyclic graphs whose nodes represent variables, and whose edges represent probabilistic relationships between variables [210]. Knowledge items, misconceptions, emotions, motivations and goals can all be represented as nodes; the Bayesian Network can then reason about observed action sequences, predict subsequent actions and analyze expected utilities. Andes is a representative ITS that uses Bayesian Networks in a physics domain [81, 374, 80].

Cognitive tutors maintain a student model, by estimating the probability that the student knows each skill in the cognitive model [8]. The student model is updated by an algorithm called *knowledge-tracing*; this algorithm is applied once for each step, where a step is a subgoal in a tutor problem. The student model update may be positive or negative, depending on whether the step was completed successfully by the student, meaning that no errors were made or hints requested.

Finally, CBM, that we discussed above as an approach to representing the domain model, can simultaneously be used as a student model: the relevance conditions of all constraints are matched to the student's solution, and so are the satisfaction conditions for all matching relevance conditions. Violated satisfaction conditions represent errors, on which the ITS provides feedback to the student. Hence, CBM does not require a separate student module or an explicit representation of student bugs [330].

The tutor model or pedagogical model can be considered as the *driving engine for the whole system* [10]. It has to decide what to teach next and how to do so. It provides feedback like a human tutor would do, by evaluating student performance, but also needs to make the interaction with the student natural.

Cognitive tutors, at least in their more common incarnation, compare each step in the student's solution to the model via the knowledge tracing algorithm mentioned above, and either flag the step (red or green for incorrect or correct), or provide a specific message (buggy message), which is attached to a mistake the student has made. If the student needs help, they can ask for a hint;

whereas some hints function as suggestions to help the student think, the student can keep asking for hints until the hint chain is exhausted, and the right answer is provided. Constraint-based tutors provide additional levels of feedback, for example, SQL-Tutor [300, 397] includes six levels of feedback to the student, including hints that are triggered by the violated constraint. In Chapter 5, we will discuss the pedagogical module in our own CBT, ChiQat-Tutor.

Since much of interaction in tutoring includes language, not surprisingly, a robust subarea of research concerns how the pedagogical module in an ITS should use natural language interaction, the topic to which we turn next.

2.4.1   Natural Language Processing (NLP) for ITSs

A conundrum that has sometimes arisen in research on human tutoring is that, in individual interactions, even inexperienced tutors are effective in increasing students' knowledge. An explanatory hypothesis is that their effectiveness stems from the conversational approach individual sessions are conducted in, as opposed to a didactic and lecture-based approach, which is not personalized to the individual student [165, 171, 46]. As we noted earlier when discussing Chi's active/constructive/interactive distinction [66], an interactive style of engagement does not mean all learning activities are interactive. However, first, a conversation can help students engage in learning activities to start with; then, by default in a conversational setting, it is easier to stir a student toward interactive learning activities.

Because of the effectiveness of human-human tutoring, starting in the early '90s with the work by Evens and colleagues on the CIRCSIM-Tutor ITS [381, 372, 111], many researchers have analyzed conversational tutoring interactions in order to first, ascertain which features of these conversations are most effective for learning and second, develop computational models of those features to deploy them in an ITS – indeed, this is the path that we have also followed, and that this book illustrates.[14]

In the earlier work in the '90s, the linguistic and pedagogical analysis of tutoring conversations was used as design guidelines for the ITS interaction model; in more recent years, machine learning, especially in terms of educational data mining, has been applied to tutoring interaction data to derive models that can then be integrated in the tutor model, or sometimes, in the fourth ITS component that directly "faces" its users, the "user interface module" [10]. Whereas the benefits of tutoring conversational interactions have often been replicated within ITS evaluation, the results are much more nuanced than the blanket assertion *Dialogic interaction is always beneficial for every student*; the results often depend on students' characteristics [418, 46], and are affected by the limitations of NLP technologies.

---

[14]Some earlier systems like GUIDON-TUTOR [75], include some natural language interaction with the user, encoded via rules, which as the authors themselves note, *work[s] rather well, although it lacks a theoretical foundation.*

A disclaimer before we start our review of ITSs endowed with NLP interaction, first in general and then specifically in the CS domain. In what follows, we include research only if it covers linguistic aspects of interaction, and/or if the speech/language interaction with the ITS is supported by NLP techniques. Many (all?) ITSs do interact with the students in a variety of ways, via menus, graphical interfaces etc, and/or provide them with feedback expressed via language, among others. Take the SQL-Tutor [300, 303, 65], which teaches students how to create queries to databases in SQL, a specific formal language, and hence falls within ITSs devoted to Computer Science. SQL-Tutor offers rich feedback in English (e.g., positive/negative feedback, worked-out examples, etc). As we discussed earlier, SQL-Tutor's pedagogical strategies are grounded in a rigorous cognitive theory, CBM. However, first, the way these strategies are modulated is not affected by a detailed analysis of human-human tutoring dialogues, and second, neither is the way they are expressed in English by the system (we are referring here to the words and syntax used to convey the system's messages). Additionally, the input to SQL-Tutor from students is an SQL query, not a question in English (in fact, in [300] the authors discuss the problems that would arise if they were to allow students to submit solutions in English); in a version of the system in which students are invited to self-explain, they are offered different alternative explanations to choose from via a menu. This should not be taken as a criticism, but as a way to circumscribe the space of the work we present here. Note that while ChiQat-Tutor does not allow students to input utterances either, it satisfies the other two tenets, being based in a detailed analysis of human-human tutoring dialogues, which affect the specific formulation of ChiQat-Tutor's interventions as well, and employing some NL generation techniques.

As noted, the group who developed the CIRCSIM-Tutor ITS, led by Evens at the Illinois Institute of Technology, was a pioneer in analyzing tutoring interactions and deploying the results of that analysis in an ITS. CIRCSIM-Tutor teaches cardiovascular physiology and uses tutoring strategies that mimic expert human tutors (who were both observed in their interactions with students, and were part of the research team). Among the issues that the Evens' group investigated are differences between face-to-face and computer-mediated tutoring sessions [381]; the functions of hinting [206], of taking initiative (being more active) on the part of the students [372, 382], of using discourse markers [229] (conjunctions such as *and, so, now* when used to connect clauses or sentences) and of deploying analogies [271, 270], and a variety of techniques related to teaching in the specific domain, including which physiological variables to teach first, and at which level of knowledge to teach [297]. CIRCSIM-Tutor was endowed with many of these strategies and several were evaluated. CIRCSIM-Tutor was shown to engender significant improvements from pre- to post-test, and was used in actual classes, which is even more notable since CIRCSIM-Tutor had to deal with severe limitations of the NLP technology available at the time [159, 111].

AutoTutor [321] is a family of ITSs, developed by a group led by Graesser at the University of Memphis and deployed in a number of different domains, from computer literacy to physics to critical thinking. Grounded in Graesser's cognitive theories of human tutoring [165, 167, 336, 170], AutoTutor's pedagogical/interaction strategies include different types of feedback (positive, negative, and neutral feedback); different ways of transferring initiative to the students (pump, prompt) and different sorts of scaffolding (hint, elaboration and splice/correction) [172, 168, 166, 169]. The AutoTutor ITSs have been found to be very effective [321]. Interestingly, to understand students' inputs, AutoTutor uses Latent Semantic Analysis (LSA) [97, 252], a vectorial representation of words in context, a precursor to the neural word embeddings of today like Glove [335] or BERT [100]. Of note is that one of the first applications of LSA had been in the context of an Intelligent Essay Assessor [121], another early success of applying NLP to educational technology, although not within a full fledged ITS.

Other noteworthy examples of early ITSs that include conversational interactions are ATLAS-ANDES and Why2-Atlas [134, 417, 362, 216, 419], which use natural language conversations to allow students to construct their own knowledge, either when solving physics problems or for writing qualitative physics essays. These efforts in the early 2000's led some of those researchers to focus, in more recent years, on a variety of issues pertaining to interactive educational technology, for example, on linguistic analysis and NLP technology to support collaborative learning and group interactions, including in MOOCs (Massive Open Online Courses) [361, 205].

Other researchers initially involved in ATLAS-ANDES & Why2-Atlas have continued to explore fine-grained approaches to providing individualized support in tutorial dialogues in ITSs. In particular, Katz, Jordan and collaborators have focused on adapting the delivery of *Knowledge Construction Dialogues (KCDs)* in Rimac, an ITS for physics they have extensively experimented with in high schools; Rimac engages students in reflective dialogue after they have solved quantitative physics problems [223, 214, 7]. A KCD is based on a main line of reasoning that it elicits from the student via a series of questions (in turn, the CIRCSIM-Tutor group were the first to study and deploy main lines of reasoning in their ITS [206]). Basically, KCDs embody an interactive script that engages students; however, the Rimac team notes that KCDs are not flexible, in that they deviate from the main path in the script only in response to a student's wrong answer: in this situation, a KCD uses a remedial sub-dialogue, and then resumes the main path in the script. In Rimac, Lines-of-Reasoning (LORs, corresponding to KCDs) are designed at different levels of granularity; they can be represented as graphs, where nodes correspond to concepts the tutor can ask questions about and arrows represent inferences a student needs to perform, to move to the next node. Once the system engages the student in a reflective dialogue, it needs to decide the level of granularity at which it will ask the next question in the LOR (or in a remediation if the previous question was answered incorrectly), to proactively adapt to the student's changing knowledge level, as reflected in the evolving student model. The system will always ask the question at the highest possible level (i.e., in the highest possible LOR) that the student can likely answer correctly.

The Rimac team compared the adaptive version of the system just sketched, with a control version where the questions in the reflective dialogue were based on pre-test and current answer to the previous question (i.e., the LOR could not change within a single reflective dialogue). The results show that students significantly learn in either condition, and there are no differences in learning between conditions. However, the experimental condition was more efficient, with a significant difference in the time students spent working with the system ($\mu = 51'26"$, $\sigma = 12'44"$ in the experimental condition, 20' faster than in the control condition). This difference held across all students, including those who came in with low prior knowledge.

Our review cannot do justice to all ITSs that use (written) NL interaction. Here, we briefly mention BEETLE-II [108], an ITS that teaches basic electricity and electronics; it supports unrestricted student natural language input in the context of a dynamically changing simulation environment, and provides contextually-appropriate feedback which is dynamically generated, and expressed via language (in one condition). Other recent examples include an ITS that supports interaction with Virtual Laboratories, such as Biotechnology, by engaging students via typed dialogue [331], and WDBT, the IBM Watson dialogue-based tutoring system architecture [4].

CIRCSIM-TUTOR, the AutoTutor systems and the others we have briefly described operate with written/typed text. Allowing students to interact with the ITS using spoken language opens a whole new area of research, including whether speech affords more opportunities for learning and/or is more effective (there is positive evidence in answer to both questions [188, 341, 267]). An ITS where the dichotomy between typed and spoken affordances was explicitly investigated is ITSPOKE [122, 123]. In ITSPOKE, spoken interaction was shown to be more effective than typed dialogue; however, unfortunately ITSPOKE did not increase student learning. An important finding was that speech recognition errors did not negatively impact learning [267].

One affordance that spoken language provides is that of conveying emotions much more directly than typed language. An important component of the research in ITSPOKE was in fact devoted to detecting students' emotions and appropriately reacting to them [265, 6]. For example in [124], these researchers found that higher disengagement was correlated with lower learning; and more specifically, that some specific types of disengagement correlated with lower learning, such as those derived from finding ITSPOKE questions too hard, or providing vague answers to game the system, i.e., to elicit information from it. We will note that in fact, there is a robust body of research within the ITS community, that looks at assessing a student's emotions during the interaction, in order to adjust the ITS intervention; a review can be found in [275]. Many investigations concern emotions detection via physiological signals (e.g., in AutoTutor [105], or in AMT [416]); from gaze [209]; or from facial expressions in ITS for Computer Science, e.g., [407] or JavaTutor (to be discussed below) [173]. However, neither does most of this research address emotions as inferred or addressed via language nor were emotions a focus of our research in ChiQat-Tutor; hence, we will not discuss this topic further.

2.4.2 Modes of Learning and ITSs

We will briefly circle back to our discussion of Modes of Learning in Section 2.1.2 to highlight how ITSs have incorporated the four we have focused on: positive and negative feedback (which we will discuss together), Worked-Out Examples and analogies.

2.4.2.1 Positive and Negative feedback

The first decades of work on ITSs focused almost exclusively on the correction of errors. This focus led to extensive analyses of subject matter, and in particular, the number of ways in which the subject matter can be misunderstood or misrepresented; in some cases, this was embodied in "bug libraries" or error inventories [48].[15] Initially, such inventories seemed like very powerful tools, and tutoring systems built around this concept were designed to support the detection and correction of errors. Experience with this type of system outside the domain of arithmetic revealed that errors varied widely from one educational context to another, and the set of 'buggy' versions of the subject matter was much larger and less precisely specifiable than it first appeared [110]. After this, the focus moved beyond either focusing only on correcting errors or on using error inventories to recognize errors; on the latter issue, earlier we already described the different ways to recognize errors, from a knowledge-tracing tutor recognizing a mismatch with the cognitive model, to a constraint-based tutor noticing violated satisfaction conditions of a constraint.

As we also noted earlier, negative feedback can be provided in a graphical fashion, e.g., via a red flag, or via an explicit notification and correction, or even via remedial subdialogues within a KCD. Many studies exist on what sort of feedback on error is more effective; one of the first was [291], which compared minimal feedback to feedback that addressed the condition that had to hold for the step to be successful but didn't, or the goal in whose context the error occurred, or both. Still to this day, many articles continue to explore the effects of feedback in educational applications [386, 402], including how to modulate it, e.g., by providing less or more targeted information to help the student overcome the impasse on their own [87, 282, 190, 123]).

In all the discussion on feedback, less attention has been paid to the role of positive feedback, in the sense we intend it here as confirmation of

---

[15] *Bug* is not quite the same as *error* [110], but we bundle them together for the purpose of this discussion.

(partial) correctness of a step, rather than as emotional support. Apart from the work by our group and some collaborators beyond what we describe in this book [88, 104, 302], Heffernan and Koedinger [194, 193] explored how to include feedback on correct student's steps even when the answer was only partially correct, on the basis of their observations of one experienced human tutor. In BEETLE-II, that we briefly touched on earlier as concerns ITSs that model tutorial dialogue, Moore and collaborators go one step beyond positive feedback, focusing on "informative tutoring feedback", i.e., *feedback that combines verification, error flagging and hints adapted to the context* [108]. In their work, *negative* and *positive* feedback are types of tutoring tactic, but so are *hedging (indicate that the student is partially correct o partially incorrect, but not explicitly.)*, and others. Whereas the authors found that different versions of BEETLE-II are highly effective, they also found that interpretation errors of the student's natural language input on the part of the system affected students' satisfaction; they provide an extensive analysis of the impact of BEETLE-II's errors but not on the type of feedback per se.

2.4.2.2 *Worked-Out Examples*

WOEs have been used in several ITSs in scientific domains, with positive but often mixed results: features of the learner and/or of the examples themselves affect either learning per se, or learning efficiency. McLaren et al. [292] integrated WOEs into a chemistry ITS that also includes tutored problem solving. No significant gains were observed over problem solving only; however, students were able to learn faster. More recently, [293] finessed the earlier study, with similar results: there were no significant learning gain differences between four conditions, including WOEs and erroneous examples, but students who studied via WOEs spent 46%–69% less time to complete the activity. Barnes et al. [269, 305] use WOEs in an open-ended data-driven logic tutor. The authors showed that WOEs were effective for novices, however had the same effect as hint-based systems. Advanced students did not benefit from the use of WOEs.

Other ITSs have used a hybrid approach. Renkl et al. [351] introduced the concept of faded examples. In this type of learning, students start using examples, and then transition to problem solving, with the system changing an increasing number of solution steps into problem-solving steps. This level of interactivity counteracts the inherently passive properties of traditional examples.

In a series of paper between 2013 and 2019, Mitrovic et al. explored adaptivity of several strategies in their SQL-Tutor that we introduced earlier, including: WOEs, supported problem solving, faded WOEs and in the most recent experiments, erroneous examples [65, 309, 310, 311, 312]. Depending on the experiment, their system alternates between the activities it offers, or it adaptively decides on the learning activity. In their most recent work [65], they first show that alternating erroneous examples to WOEs engenders more learning than providing only WOEs, and that an adaptive strategy that provides students with either a WOE or an erroneous example based on their performance results in comparable but more efficient learning.

2.4.2.3 Analogy

Not many ITSs include analogy as a pedagogical strategy, although theories of analogy (or similarity) may affect the ITS from different points of view; for example, in [440], WOEs analogous to a subtraction problem solved incorrectly by the student, are generated automatically in order to address the student's mistakes. As far as we know, the first ITS to include analogy as a pedagogical strategy was the Bridging Analogies tutor [308], which deployed intuitive physical scenarios to explain less intuitive concepts, in order to address students misconceptions in qualitative Newtonian physics – for example, if the students answered to a question regarding the forces acting on a book laying on a table saying that there are no forces, the ITS would invite the student to think about holding the book in his/her hand.[16] A formative evaluation of the ITS was conducted with 15 subjects, all of whom started the session with a wrong answer to the target situation (nine of the 15 further indicated high confidence in that original misconception, and the rest, low confidence). By the end of the intervention, all but one had not only the correct answer, but high confidence in the correct answer, and they explained it correctly. Importantly, none of the subjects indicated they had experienced frustration while using the ITS.

In the context of the CIRCSIM-Tutor project we described earlier, Lulis studied how their expert tutors employed analogies in teaching cardiovascular physiology [271], and then implemented them in the system [270]. In the CIRCSIM tutoring dialogues, human tutors used analogies 51 times, of which 29 with another neural variables, and the others with a smattering of concepts from different domains, such as balloons, brakes, etc. She also found that about one-fifth of the analogies were used after a correct inference on the part of the students, to enhance their understanding, and the rest, to correct incorrect inferences. [270] describes how analogies were implemented in the system, paying particular attention to their integration in the discourse planner that CIRCSIM-Tutor was using at the time. As far as we know, no formal evaluation of the version of Circsim-Tutor including analogies was conducted.

To conclude, [319] analyzes the respective effectiveness of worked-out examples by themselves, accompanied by self- explanation or used as analogies, in the sense that after working with the same WOEs as in the control condition, students would compare and contrast two further examples. The experiments were conducted in the Physics 1 course at the US Naval Academy. The self-explanation and analogical condition were effective for high learners on

---

[16]It is questionable whether this qualifies as an analogy per se. There are aspects of the situations the authors relate that fall more directly under an analogy framework; for example, whether your hand is more volitional than a table, and as flexible as a spring.

conceptual learning, while results on problem solving were mixed. Similar results were obtained with the same materials and the same type of subjects but with the students working in dyads rather than individually [138]. These experiments were conducted in the classroom and not with an ITS; however, this work is relevant to some of the comparisons we will discuss in Section 5.3.1, where one WOE condition will be augmented with an initial analogy.

2.5   ITSS FOR COMPUTER SCIENCE AND NLP

In the space of NLP for ITSs, we are in particular interested in those ITSs that operate in a Computer Science domain, like ChiQat-Tutor. Not too many fall in this specific class, even more so, given our focus on computational problem solving, not definitional knowledge. For example, initially, AutoTutor was applied in the realm of Computer Science literacy, specifically, factual knowledge about hardware and software (for example, *what is a CPU?*); computational thinking was not a main focus for that incarnation of AutoTutor. However, we will briefly review ITSs for CS in general, before we discuss those few ITS for CS which are based in language data and/or are using NLP (our focus, as we mentioned earlier).

2.5.1   ITSs for CS

To start our overview of ITSs for CS, we observe the following paradox: ITS and more in general, AI for education originates in CS itself; however, other than few pioneer ITSs some of which we have already touched on (AutoTutor, SQL-Tutor) or we will discuss below (the LISP Tutor, ProPL), the space of ITSs for CS was very sparse up to the last 10 years or so. This probably depended on the popularity of CS itself as a field of study: for many years, CS was not a major in high demand, and as we noted earlier, enrollment in CS crashed in the early 2000s. Currently, given the great demand for CS education, and initiatives like *CS for all* that we mentioned earlier, the need to support students studying CS at different stages of schooling has increased dramatically, and so has the range of AI technologies devoted to CS education.

Before discussing more recent developments, we should mention another pioneer ITS in CS, the LISP Tutor, a very early Cognitive Tutor [349, 16, 83]. The Lisp Tutor was developed mainly *to shed light on issues concerning the nature of cognition* [16, p. 468]. At the same time, the LISP Tutor was used to teach LISP programming at Carnegie Mellon University for several years, by providing students with programming exercises and giving help as needed. The tutor produced significant performance improvements, especially in terms of efficiency: in two studies, students completed the exercises in one-third to three-quarters of the time taken by control subjects who completed the exercises on their own. In general, students in the experimental and control conditions scored equally well on tests, and in one of the studies, experimental subjects scored about one letter grade higher on the final test. In the

LISP Tutor, feedback and remediation were provided in accordance to how cognitive tutors use production rules to model procedural knowledge, and knowledge tracing to relate students' solutions to correct solutions (please see Section 2.4). The LISP Tutor was also used to explore different styles of providing feedback. For example, in the *demand-feedback* condition, feedback was provided when students requested it, rather than immediately when recognizing an error – especially advanced students had noted that they wanted to have time to explore the code and that they may be able to self-correct. Not surprisingly, in the demand feedback condition, students took longer to solve problems; on the other hand, the LISP Tutor caught about a third of an error less per exercise than in the immediate feedback condition, thus confirming students' claims that they would be able to correct some errors themselves.

Moving now to current research, [31] provides a recent overview of AI-Supported Education in Computer Science. Interestingly, all the papers focus on college students, even if some of these interventions could be used at the high school level, especially in AP (Advanced Placement) courses. Apart from the two papers on AI-supported dialogue based systems that we will discuss below (JavaTutor [429] and our own KSC-PaL [203]), the other 6 papers cover the gamut from support for programming in Python [359] or Haskell [155], to investigating affect for novice programmers [42]; from aiding in logic reasoning and problem solving [304], to supporting a remote laboratory [47], to focusing on search algorithms [176]. All papers are data driven in some sense, and many do focus on the feedback and hints to be provided to the student, even if they do not focus on the language/NLP aspect of such feedback or hints (strictly speaking, [42] and even more so, [47] are not ITSs, but the issues they describe are relevant to ITSs nonetheless).

Whereas programming is not the only topic CS students need help with, many ITSs focus on it, especially on helping novice programmers. Apart from the two ITSs mentioned above for Python and Haskell programming respectively, many more exist, with different flavors and topics; a recent overview of 14 ITSs on programming can be found in [91] (they include our own ChiQat-Tutor as applied to recursion, even if our focus in not on programming per se, see Section 6.2). This overview focuses on three questions specifically: the programming languages that are being taught; the types of supplementary features used in the ITS and how they are implemented (e.g., pedagogical strategies, but also visualizations or reference materials) and the parts of the tutoring process that are adaptive within the ITS. They conclude (perhaps not surprisingly) that *It is evident that there is no standard combination of features that have been utilized within the field of intelligent tutoring applied to programming education.*

One of the ITSs mentioned in [91] belongs to the suite of web tutors, called Problets, developed by Amruth Kumar and his group in the last 20 years. The Problets tutors focus on a broad range of concepts pertinent to introductory CS and programming, including parameter passing, loops, expression evaluation [239, 383, 94]. Kumar's research program includes contributions to every facet of an ITS: from the automatic generation of problems, to assessment, to pedagogical strategies and the careful evaluation of their modulation, to long-term retention of concepts [244, 246, 247, 248]. Notably, every new incarnation of a Problets tutor or variation thereof is always evaluated in a real classroom or laboratory [240, 242]. Kumar's group has also explored a range of broader questions regarding using Problets, and more in general, software tutors in CS, including the effect of tutors on self-efficacy [245], and the impact of demographics and stereotype threats on using online software tutors [241, 243, 249].

Finally, especially because of the current focus on data driven methods, there are a number of recent papers that extract patterns from logs of CS students activities [29, 276, 366, 20, 283]; whereas these results are not necessarily incorporated in ITS for CS as of now, they are bound to affect some aspects of CS ITSs in the not so distant future. Indeed as we will describe in Section 4.5, we were early adopters of a similar approach to discovering patterns of behavior in order to modulate feedback in ChiQat-Tutor.

2.5.1.1 NLP in ITSs for CS

We will now discuss the subset of ITSs for CS that are relevant to our focus as described earlier: they are grounded in human-human tutoring interactions, or they use NLP technology either in understanding students, or in providing feedback to them.

An ITS from the early 2000s that used dialogue to engage students is ProPL [253, 254]. In this case, students are novice programmers and ProPL focuses on helping them develop better problem solving and planning skills. Expert programmers are thought 1: to possess schemas, i.e., *reusable "chunks" of knowledge representing solution patterns that achieve different kinds of goals* [390] and 2: to be able to correctly invoke and combine them to solve the problem at hand. Novices are lacking especially as concerns the second point. [253] collected a corpus of human-human tutoring dialogues in this domain, and discovered a three-step pattern, which includes progression from a goal to a general description of how to achieve it (a schema), and ultimately to a plan, i.e., an implemented schema. Human tutors ask questions at each step to elicit knowledge from novices, and to help them build such knowledge if lacking.

After presenting a new problem to a student in a dedicated window in a four window interface, ProPL starts a dialogue with the student (in a second dedicated window). ProPL first checks that the student has a basic understanding of what the program should do (i.e., what its inputs and outputs are). Then, it engages the student according to the three-step pattern just described. First, ProPL asks a student about a programming goal, which once identified, is posted in the design notes pane; a "how" question is then asked, and if a student answers correctly, the schema is added under the goal as a comment; then the student is asked about their plan, i.e., how to pull the various schemas together at the end, when a correct solution is reached, the pseudocode window is updated. Of course, at any point, the student may answer incorrectly, or simply type "I don't know". Then, a subdialogue starts to correct the mistake. ProPL uses the Atlas tutorial dialogue planning system [134, 417] we mentioned earlier, and specifically, the concept of *Knowledge Construction Dialogue (KCD)*, which we also discussed earlier.

ProPL was evaluated in a small study. Volunteers were recruited among students enrolled in an introductory programming class using Java or C. Two conditions were run, one with ProPL as just described and one with the same interface, but with the interactive dialogue substituted with canned text, deliberately formulated to mirror the information that the full ProPL system provides. The results showed that students in the experimental condition appeared to have acquired deeper skills at schema composition and planning; for example, they omitted fewer plan parts than students in the control group, and worked more at the level of schemas and plans rather than at the line-by-line level typical of novices. However, there were no differences in performance in solving the decomposition problem on a written test.

OSCAR-CITS [257, 256, 90] focuses on a conversational ITS (CITS) that can predict and adapt to the learning style of the student via natural language dialogue; the topic is SQL, the database query language we previously mentioned. The authors present a step-by-step approach to developing such a CITS. They do not focus on naturally occurring dialogic interactions between tutors and students; rather, their goal is to replicate the effectiveness of *tutorials*, which, in some British universities, are weekly sessions with a tutor and a small group of students (from one to six) where students receive direct feedback on their assignments and are expected to actively participate in discussion. Additionally, a focus of these researchers is to capture personal learning styles (as sets of behavioral variables such as *Dislikes surprises* or *Remembers what s/he sees*) via their language proxies, and to personalize the interaction accordingly. To develop their CITS architecture, Latham, Crockett and colleagues interviewed tutors *to identify important SQL concepts for the tutorial syllabus*. They identified ten tutorial questions and a multiple choice question test to cover the learning outcomes of the tutorial [257, p.99]. Starting from these questions, and based on studies of interactions in tutorials, the authors designed a three level (social, tutoring, and discussion) conversational structure. Importantly, they mapped each of the tutorial questions to generic question styles and templates that depend on the behavioral variables they had identified. Finally, the authors manually created scripts to actually conduct the dialogues. In several experiments, they showed that OSCAR-CITS does help students learn, and can adapt to individual learning styles. The main difference between our approach and theirs is that they do not directly model human - human tutoring interactions; additionally, the development of the conversational behavior of the CITS is not automatized in any way, even if in their more recent work [90] they automatically learned fuzzy decision trees to link behavioral variables to specific learning styles.

A very compelling ITS that uses NLP for CS is JavaTutor, initially developed at North Carolina State University [44, 45, 113, 428, 115]. A large part of the research focused on linguistic analysis and computational modeling of human - human tutoring dialogues in an introductory programming course, in particular as concerns the dialogue acts underlying the participants' utterances. An inventory of dialogue acts was developed, grounded in both inventories of general dialogue acts [394] and inventories specific to tutoring dialogues [266]. On the basis of this coding scheme, the 48 dialogues that had been collected were annotated; those 48 dialogues included approximately 4800 utterances, of which roughly two-thirds due to the tutors. Additionally, all student task actions were captured in a parallel event stream, and annotated for action type (i.e., *declare new array*) and for correctness. Next, a series of papers was devoted to computationally modeling these dialogues, via Hidden-Markov Models (HMM) and later via clustering. Initially, in [44, 45], a set of hidden dialogue states was automatically learned, and found to correspond to tutoring dialogue modes (learning was carried out via the standard Baum–Welch algorithm to learn HMMs). Once the HMM was derived, another standard algorithm for traversing HMM, the Viterbi algorithm, was used to classify the student's utterances in context. Subsequently [115], clustering methods were used to achieve unsupervised classification of dialogue acts, taking into account task and dialogue history features, and later, posture and gesture information as well. Finally, these findings were used to drive the interaction between students and JavaTutor, an ITS that would help them with the basics of Java programming. Supervised and unsupervised models of classifying dialogue acts were evaluated in JavaTutor [114], and in particular, the unsupervised model was shown to adequately support students' learning. However, as far as we know, no large evaluation of JavaTutor was conducted.

S.T. is a recent Socratic ITS (hence the name, from *Socratic Tutor*) by Rus and colleagues [367, 401, 12]. S.T focuses on source code comprehension, and is web-based. Its focus is on helping novice programmers better understand programming concepts. The authors claim that S.T. is programming language independent and can be used to teach any programming language; their current experiments focus on Java. As in the traditional ITS architecture we described earlier, S.T. is comprised of a student model, a domain model and an interaction model. The latter is defined by the dialogue policy which selects the next dialogue move, which depends on the output of the Natural Language Understanding module which compares the student response to an ideal response.

The focus of the S.T. ITS is on code understanding and output prediction. Hence, the S.T presents students with code snippets in one window and asks the student to explain the code in writing and predict its output, in a separate window (a third window shows the dialogue history) Then, S.T. guides the student through a socratic dialogue, i.e. it asks questions that *were manually designed by experts following the Socratic method's guidelines* [12]. S.T. also aims at detecting misconceptions.

The S.T system provides support to learners in the form of hints in two cases: (1) when a student asks explicitly for help or (2) when the answer is incomplete or incorrect. In both the cases, the S.T starts a 3-level feedback loop. At levels 1 and 2, S.T. provides brief hints and explanations, and presents the student with a fill-in-the-blank question (which is a hint in itself). At level 3, S.T. provides a multiple choice question, and if the answer is still incorrect, then the solution is presented and S.T. moves to the next code snippet.

S.T. focus on self-explanation is motivated by much previous research on the effectiveness of self-explanations [67, 68], and by their own study [401]. 26 students were presented with 4 code examples in Java (in an online interface but not the full S.T.). For each such code example, participants were prompted to predict its output and explain their thinking in writing. Then, they were shown another 4 code examples for which they were simply asked to predict the output (no explanation, just prediction), on which their learning was assessed, in addition to a traditional pre-/post-test. The participants were divided into two conditions according to the order in which they engaged in self-explanation and prediction: Self-Explanation First and Prediction First. The findings showed the importance of Self-Explanation: participants in the Self-Explanation First group performed 31% better, than participants in the Prediction First group, as concerns the prediction score. Additionally, there was also a significant difference in mean learning gains between the two groups (as adjusted for pretest scores), with a small effect size (0.20). Finally, there was a strong, positive correlation between the number of content words in the self-explanations and learning gains.

The S.T. system per se has been evaluated with 34 students in an undergraduate introductory CS class; students worked on nine Java snippets with the system, and took a pre-test, and an identical post-test. The students significantly learned; when the students were divided into two groups based on mean pre-test score, it became apparent that the bottom half of students learned significantly more. Further analysis of the effect of the three-level feedback showed improvement in correctness of answers, but no significant results were reported.

As per the authors, the S.T. system and attendant studies are going to be the focus of further development and investigation, including investigating other common instructional strategies in the ITS, and comparing written to spoken self-explanations.

Moving now to our own work, apart from ChiQat-Tutor that of course is the main topic of this book, we explored a different angle pertaining to collaboration, how it evolves in conversations between peers, and how the same ChiQat-Tutor can be adapted to act as a peer [226, 227, 203, 202]. Briefly, we started from cognitive underpinnings that distinguish productive peer interactions in a learning setting, and specifically, the notion of Knowledge Co-Construction (KCC) [61, 189, 306, 93]. A KCC episode is a series of utterances and actions in which students are jointly constructing a shared meaning of a concept required for problem solving. However, from a computational modeling point of view, a KCC episode is very hard to recognize; hence, we set out to find easier correlates of KCCs. Linguistic analyses of collaborative dialogues [425, 215, 177, 74, 382, 85, 192, 438, 320] show that the concept of *initiative* embodies the interlocutors' participation in the dialogue and is indicative of their contributions to the problem-solving activities they engage in during the dialogue. Informally, a speaker takes initiative when she contributes content which is new with respect to what her dialogue partner suggested and is not invited by her partner – for example, most of the time asking an unsolicited question shows initiative, whereas answering to a question does not. Further, initiative can be distinguished as *dialogue* and *task* initiative, depending on which realm of the interaction the contribution falls in.[17] Finally, if one participant dominates the conversation and/or the problem solving activities, initiative will not shift between speakers as much as when the participants are equal contributors.

We collected a corpus of 15 computer-mediated peer interactions in the same domain of introductory data structure (linked lists, stacks, binary search trees) that ChiQat-Tutor focuses on. We annotated these interactions for KCC episodes, and for dialogue and task initiative. We found that task initiative shifts occur more frequently within KCC episodes than outside these episodes. Additionally, we found that task initiative shifts within KCC episodes correlate with learning for low pre-testers, and total task initiative shifts correlate with learning for high pre-testers. On the basis of these findings, we designed KSC-PaL, a software agent that could collaborate with students to solve problems on data structures. KSC-PaL used the same architecture of ChiQat-Tutor-v1 but added a dialogue planner to support the automatic tracking and shifting of task initiative. KSC-PaL was able to recognize whether the student has task initiative given their utterance or their action, using features such as dialogue initiative, the dialogue act of the previous utterance, and the knowledge score of the student, computed on the basis of the student model; the accuracy is about 72% on utterances and 87% on actions. On the basis of this, KSC-PaL was able to assess whether the student was active enough in the problem-solving situation by keeping track of the average level of task initiative shifts based on what had transpired in the dialogue so far; if this level fell under a certain threshold, KSC-PaL would (indirectly) invite him/her to take initiative by using one of six moves (that we dubbed "shifters"). Shifters included using hedges (mitigating devices such as *could* in *that could be wrong*), asking for feedback, issuing prompts (short, non-propositional utterances such as *yeah, hmm*) or even issuing wrong statements: these six moves also originated from the analysis of our dialogues.

---

[17]Note that the two types of initiative do not necessarily coincide: for example speaker 1 could ask a question showing dialogue initiative, and holding on to dialogue initiative when speaker 2 answers; however, speaker 2's answer could advance problem solving and hence show task initiative, as in *Sp. 1: Do you know how to declare an integer variable score? Sp. 2: Use* `int score`.

KSC-PaL was evaluated in a small study, where the two conditions differed in whether KSC-PaL was endowed with the ability to track task initiative shifts, and consequently, with the ability to encourage the student to take the initiative. Whereas students in both conditions learned significantly, there was no difference in learning by condition. We then examined task initiative shifts specifically. We showed that the experimental version of KSC-PaL did produce significantly more "shifters" (note that the control condition can produce some of those moves too, since they are regular dialogue moves); that there was a significant difference in students taking the initiative in the experimental condition; and we did find that task initiative shifts had a small but significant effect on post-test score, after regressing pre-test score out.

II

From Human Tutoring to ChiQat-Tutor

CHAPTER 3

Human Tutoring Dialogues and their Analysis

With Stellan Ohlsson
University of Illinois at Chicago

With Mehrdad Alizadeh
Lexis Nexis

With Lin Chen
Cambia Health Solutions

With Rachel Harsley
Google

CONTENTS

N ATURALISTIC data on human–human tutoring in introductory Computer Science affords us both cognitive and design insights. In this chapter, we will discuss the entire process we followed as concerns corpus collection and analysis. We collected 54 human–human tutoring dialogues, that we transcribed and annotated for features of interest, both at the individual utterance level and at the segment level; the annotations were validated via assessment of intercoder agreement. We then performed multiple regression analysis that uncovered which features of the dialogues correlate with learning. The analysis is motivated both theoretically, to gain insight into which pedagogical strategies are more effective for learning, and practically, since the design of the ChiQat-Tutor ITS will focus on computationally modeling those effective features.

3.1 DATA COLLECTION

Our domain of interest is Computer Science (CS), specifically, introductory data structures such as *linked lists, stacks* and *binary search trees* and the algorithms that manipulate them. Appendix A contains a primer on these data structures for readers unfamiliar with them; two illustrative figures are included in Figures 3.1 and 3.2, respectively; no figure is included for stacks, you can think of stacks as trays in a cafeteria (the BST figure 3.2 illustrates the BST ordering property: the items are ordered alphabetically, with all the items preceding *milk* on the left, and the ones following *milk*, on the right).

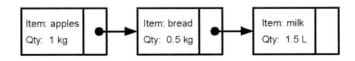

Figure 3.1 A linked list containing a grocery list.

We collected a corpus of 54 one-on-one tutoring sessions on these three data structures. Each individual student participated in only one tutoring session, with a tutor randomly assigned from a pool of two tutors. One of the tutors was

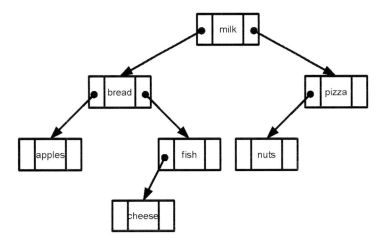

Figure 3.2 A binary search tree containing a grocery list

an experienced CS professor, with more than 30 years of teaching experience, including one-on-one tutoring. The other tutor was a senior undergraduate student in CS, with only one semester of previous tutoring experience.

Students took a pre-test right before the tutoring session, and an identical post-test immediately after. The test had two problems on lists, two problems on stacks and four problems on trees. The first problem on lists, the first problem on stacks and three problems on trees asked the student to manipulate the data structure in order to execute a certain task; the second problem on lists was a debugging problem (namely, the student was asked to find the error in a snippet of code); the final problems on stacks and trees were both verbal explanation problems. Each problem was graded out of 5 points, for a possible maximum score of 10 points each for lists and stacks, and 20 points for trees. Pre- and post-test scores for each topic were later normalized to the [0..1] interval. The reader can find the full pre-/post-test in Appendix B.

The tutors did not know the problems on the pre-test, since we did not want them to tutor to the test, but they were presented with a high-level summary of the pre-test results. Additionally, they had a predefined set of problems they could use during tutoring, and were alerted when the tutoring session had lasted 45 minutes. Even so, tutors had considerable latitude in how long they tutored students. In a few cases, they skipped topics altogether; in others, they decided to continue beyond 45 minutes, while yet in other cases they stopped tutoring around 35 minutes, hence the variability in the length of the tutoring sessions (see Table 3.2 below). In both the conditions, the vast majority of students were recruited from our undergraduate introductory CS classes, before they studied the data structures of interest; there were also few graduate students from other engineering departments, but none of them were familiar with the data structures in the pre-test, as the learning results show.

An additional group of 53 students who were taking an introductory data structure class served as controls during one class period. They took the pre-test at the beginning of class, and the post-test at the end of class. In between the two tests, instead of participating in a tutoring session they attended a 40-minute lecture about unrelated mathematical concepts[1] which were taught alongside elementary data structures in this class. The rationale for such a control condition was to assess whether simply by practice effect (taking the test on the topic, and in fact, repeating it twice), students would learn about data-structures. We had actually also run a reading control condition, where 28 students read relevant sections of the textbook between the two tests. However, in retrospect, the students in this reading condition were too advanced in that they had already seen linked lists and stacks in a prerequisite class. Not surprisingly then, we found no significant difference in learning gains between the human tutored conditions and the students in the reading control condition. However, we found a significant difference in learning gains between these two conditions as concerns binary search trees ($F(1; 81) = 7.0, P < 0.05$), which these students had not yet seen. Since it is well established that human tutors are more effective than conditions such as reading relevant material in a textbook [111, 418, 21], we decided not to rerun the reading condition with a more appropriate group of students, and to rely on the more robust alternative control condition we just discussed. Hence, the reading condition will not be discussed further.

3.1.1 Learning Outcomes in Human Tutoring

If there were no learning among our tutored subjects, it would obviously not make sense to annotate the corpus to discover how students learn. Hence, the first finding we report is that the tutored students did learn, whereas those in the control group did not. Before delving into our results, we will discuss the measure of learning gain we used throughout our work, starting with the human data collection but also later when evaluating various versions of ChiQat-Tutor.

3.1.2 Measuring Learning Gains

The simplest way to calculate learning gain is via *absolute learning gain*, which is the delta between a student's pre-test and post-test scores; this is in fact the measure we adopted. With this calculation, all gains are considered equal, i.e., all students who score an extra point on the post-test will have an equal learning gain. It can be calculated using the formula in 3.1.

$$\langle g \rangle = \langle post \rangle - \langle pre \rangle \tag{3.1}$$

An issue with absolute learning gain arises when samples are collected

---

[1]The lecture was about discrete math.

across multiple student populations where pre-test scores wildly vary. Hake et al. devised an alternative measure called *normalized learning gain*, whereby learning gains are relative to how much a student can gain after intervention [179]. It can be calculated using the formula in 3.2.

$$\langle g \rangle = \frac{\langle post \rangle - \langle pre \rangle}{1 - \langle pre \rangle} \tag{3.2}$$

However, normalized gain does have some fundamental flaws, such as not being able to cope with negative learning. *Normalized change*, introduced by Marx and Cummings [281], improves upon normalized gain by taking into account negative gains; additionally, in this computation those subjects are dropped, who either gain a perfect score, or zero, in *both* pre- and post-test (captured by *c=drop* in Equation 3.3). Normalized change can be calculated with the formula in 3.3.

$$c = \begin{cases} \frac{\langle post \rangle - \langle pre \rangle}{1 - \langle pre \rangle} & post > pre \\ drop & post = pre = (1 \text{ or } 0) \\ 0 & post = pre \\ \frac{\langle post \rangle - \langle pre \rangle}{\langle pre \rangle} & post < pre \end{cases} \tag{3.3}$$

Using a more sophisticated learning gain method may not always be best. Either of the normalized variants gives a far greater weight to those students who scored very highly in the pre-test. The key to deciding whether absolute learning gain is sufficient depends on how much the pre-score of students varies between conditions. If pre-scores do vary considerably, then normalized variants would be more suitable than absolute gain. In our experiments, in the vast majority of cases we found no significant differences in pre-test scores. On the rare occasions in which we found them, they highlight population differences such as, course students are enrolled in, and hence, background they are assumed to have. In those cases, specifically in the evaluation of ChiQat-Tutor-v2, we evaluate effects on those groups of students separately, and hence, using absolute learning gain is appropriate.

3.1.3 Learning Effects

Paired samples t-tests revealed that post-test scores are *significantly higher* than pre-test scores in the two tutored conditions for all the topics, except for lists with the less experienced tutor, where the difference is only marginally significant. If the two tutored groups are aggregated, there is significant difference for all the topics. Students in the control group achieved a modest but significant gain ($t(52) = 2.2181$, p-value $= 0.03094$); however, they show significant learning only for stacks, but not for lists and trees. Means, standard deviations and t-test statistic values are reported in Table 3.1. N represents the number of subjects in that specific group (for stacks, N is lower for the

Table 3.1 Learning Gains and t-Test Statistics

Topic	Tutor	N	μ	σ	t	p
List	Novice	24	.09	.22	-2.00	.057
	Experienced	30	.18	.26	-3.85	< .01
	Combined	54	.14	.25	-4.24	< .01
	Control	53	.01	.15	-0.56	ns
Stack	Novice	24	.35	.25	-6.90	< .01
	Experienced	24	.27	.22	-6.15	< .01
	Combined	48	.31	.24	-9.20	< .01
	Control	53	.05	.17	-2.15	< .05
Tree	Novice	24	.33	.26	-6.13	< .01
	Experienced	30	.29	.23	-6.84	< .01
	Combined	54	.30	.24	-9.23	< .01
	Control	53	.04	.16	-1.78	ns

Experienced tutor, because the tests administered to the first 6 subjects in our data collection did not include problems on stacks).

The learning gain, expressed as the difference between normalized post- and pre-test scores, of students that received tutoring is *significantly higher* than the learning gain of the students in the control group, for all the topics. This is showed by ANOVA between the aggregated group of tutored students and the control group. For lists, $F(1, 106) = 11.0$, $p < 0.01$. For stacks, $F(1, 100) = 41.4$, $p < 0.01$. For trees, $F(1, 106) = 43.9$, $p < 0.01$.

We also observe that lists appear more difficult for students to learn than stacks or trees. Indeed Table 3.1 shows that in the CS tutoring corpus the average learning gain is only .14 for lists, but .31 for stacks and .30 for trees; whereas students have the lowest pre-test score on stacks, and hence, they have more opportunities for learning, they learn as much for trees, but not for lists. We hypothesize that lists may be difficult because they are the first recursive data structure that students learn, and recursion is a notoriously difficult concept to grasp, as we will discuss in Section 6.2. Additionally, students must develop the ability to move seamlessly among multiple presentations of the same concept, and graphical representations for trees may be better known since they are used outside of CS (e.g., genealogical trees).

There is *no significant difference* between the two tutored conditions in terms of learning gain. Since students did not learn more with the experienced tutor, we decided to pursue a correlational rather than a contrastive approach: we investigated what happens in the tutoring sessions independent of tutors rather than what tutors do differently from each other. Whereas specific learning outcomes may be idiosyncratic to specific tutors, our specific

Table 3.2 Duration of Sessions in the CS Tutoring Corpus

Topic	N	Length (minutes)				Total length (hours)
		Min	Max	μ	σ	
List	52	3.4	41.4	14.4	5.8	12h 30'
Stack	46	0.3	9.4	5.8	1.8	4h 24'
Tree	53	9.1	40.0	19.2	6.6	16h 57'
Total Sessions	54	12.8	61.1	37.6	6.1	33h 51'

interest was in uncovering which theoretically motivated pedagogical strategies they may adopt.

One final comment with respect to the validity of our pre-/post-test. We developed the questions in order to assess understanding of the targeted data structures, not just superficial memorization. The authors of this book have considerable experience in teaching these data structures; hence, to craft these questions, we could draw on textbook examples and on previous exams we had administered. In conclusion, we consider these tests to be valid measures of the targeted material, as are the problems that students were asked to solve in the ITSs that we subsequently developed.

3.2 TRANSCRIPTION AND ANNOTATION

The 54 tutoring sessions were videotaped; the cumulative session length amounts to a total of 33 hours and 51 minutes. More detailed statistics on duration (by session and by topic) are presented in Table 3.2. In this table, *session* refers to the full interaction between a tutor and a specific student; as the reader will notice, not all sessions covered all topics. As we noted earlier, tutors were free to skip a topic if the high-level summary of the pre-test indicated the student knew the topic well enough already; this happens in particular for stacks, which were covered 46 times out of 54 sessions, i.e., 85% of the times, whereas lists and trees were almost always discussed.

The videotaped material was transcribed following a small, relevant subset of conventions from CHAT, the transcription manual of the CHILDES project [272].[2] An utterance is a natural unit of speech bounded by breaths or pauses. Figures 3.3 through 3.7 show excerpts from our dialogues, which include some of the transcription conventions. These excerpts are annotated with the codes we will describe in the following sections; **T** is the Tutor, and **S**, the Student.

As concerns transcription, for example, individual letters pronounced as such, (eg. *q*), are indicated with q@l (many examples throughout the 5 excerpts, since our data structures contain either letters or numbers as content). The string +... (lines 216, 219, 227 in Figure 3.3) marks trailing; correspondingly, the string ++ on the line following trailing (as in students' turns 217, 220

---

[2]Available at https://talkbank.org/manuals/CHAT.pdf, latest update August 2020.

Table 3.3 Turns, Utterances and Words Across Tutors and Students

Unit	Tutors		Students	
	μ	σ	μ	σ
Turns	46.6	37.4	45.5	37.3
Utterances	186.0	107.6	48.6	40.0
Words	1971.8	1072	155.7	151.3

and 228) marks completion (technically, *latching*). The string +// (line 219 in Figure 3.3) indicates that the speaker is self interrupting.

Angle brackets mark repeated speech in some cases; in others, abandoned speech. If a stretch of abandoned speech is followed by [//], this indicates *retracing*, namely, *when a speaker starts to say something, stops, repeats the basic phrase, changes the syntax but maintains the same idea* (see for example lines 13 and 14 in Figure 3.4). If angle brackets occur in two adjacent utterances by two different speakers, such as lines 30 and 31, and 33 and 34, in Figure 3.4, they represent overlap between the two speakers. The diacritic [>] at the end of the first speaker's utterance indicates *overlap follows*, and likewise, the diacritic [<] at the beginning of the second speaker's utterance, *overlap precedes*.

Pauses are also marked with diacritics, for example, # indicates a short pause (lines 538 and 545 in Figure 3.5); finally, 'xxx' marks unintelligible speech, for example, at the end of line 17 in Figure 3.4.

Table 3.3 shows the average number of turns, utterances and words, per tutor (across the two tutors), and per student. Tutors are shown as an aggregate since there are no significant differences between these two tutors along those dimensions; and as we mentioned earlier, there are no significant differences in learning gains between the two tutors. These numbers show that tutors talk much more than students, producing more than 10 times as many words. Tutors produce longer turns, consisting of 4 utterances on average, while students' turns contain 1 utterance on average, and tutors' utterances are longer, consisting of 10.6 words on average, while students' utterances contain 3.2 words on average. Note that the number of tutor and student turns is equivalent by construction, since a turn is an uninterrupted sequence of utterances by one of the speakers.

The verbosity of our tutors may surprise the reader. They talk a lot, to the tune of producing 93.5% of the total words! Certainly, they do not resemble an idealized Socratic tutor, who is supposed to prompt the student to construct knowledge by themselves [68]: in that scenario, students should do most of the talking. In our view, this is an idealized view of the tutoring process; indeed, expert tutors are not a panacea. For example, the small study described in [71] found that their expert tutor did not effectively adapt to the bottom half of the tutees. Going back to interactivity, whereas our tutors may be extreme in their verbosity, even tutors who appear to come closer to the ideal do talk more and with longer sentences than students. In the CIRCSIM face-to-face studies, tutors produced 63% of the words [111]; their utterances were also longer than students' utterances, with 8.2 words per sentence, as opposed to 5.8 words from students. Person's expert tutors [55] also talk more than their students, producing 77% of total words (p.c.). An even more telling result is from [196]: they studied 46 tutors tutoring concepts in biology, and showed that students' deep learning was positively affected by a tutor providing more scaffolding but less explaining, which they define as *tutor's interactivity*. However, they found that expert tutors (21 biology teachers) were not more interactive than the 25 college students of biology (although we should point out, that the pedagogical expertise concerned experience in the profession, not experience in one-on-one tutoring). [196] concludes:

> Our findings suggest that a more complete understanding of inter-activity in tutoring requires a differentiated approach considering interactivity as a multifaceted phenomenon.

We could not agree more; in fact, we argue[3] that it is the notion of expert tutor itself that is distracting, and that rather than measuring tutors' expertise, we should focus on a correlational analysis that correlates what tutors and tutees do in a tutoring dialogue, with learning gains. This is in fact the analysis we present in this chapter.

3.2.1 Annotation

Our annotation follows directly from the Multiple Modes Principle, and from the pragmatics of dialogue acts, both of which we outlined in Chapter 2. As we discussed in Section 2.1.2, we decided to focus on the following six tutoring moves, that address six of the original nine modes of learning. From the tutor's point of view, he or she can provide the student with:

1. Procedural information on how to perform the task (for the learner to operationalize that advice)

2. Declarative knowledge about the domain (for the learner to reason from it)

3. Positive feedback, to confirm that a tentative step is in fact correct

4. Negative feedback, to help a student detect and correct an error

5. A worked-out example/demonstration

6. An analogy

Our focus on these six moves was informally confirmed by a read through the

---

[3]An argument that we started putting forward in 2007 [329].

entire corpus, which we examined for impressions and trends, as suggested by [70].

We translated the first four types of tutorial interventions just listed into the following Dialogue Acts (DAs): Direct Procedural Instruction (DPI), Direct Declarative Instruction (DDI), Positive Feedback (FB+) and Negative Feedback (FB-). Besides those four categories, we additionally annotated the corpus for Prompt (PT), since our tutors did explicitly invite students to be active in the interaction. Types 5 and 6 (worked-out examples and analogies) pertain to annotation of meaningful episodes, and will be discussed later.

About half of these annotations apply to individual utterances, namely, DDI, DPI and PT; FB+ and FB- are defined as short sequences of utterances, whereas worked-out examples and analogies may cover long stretches of the dialogues. Examples are provided throughout, and can also be found in Figures 3.3 and 3.4.

Direct Procedural Instruction (DPI)   occurs when the tutor directly tells the student what task to perform. More specifically, utterances are marked as DPI if:

- they mention correct steps that are part of the solution of a problem *(It doesn't matter where you part point F xxx just point at that node)*

- they describe high-level steps or subgoals *(So let me show you how to find a node that has b@l in it.)*

- they describe strategies and tactics *(Well alright so that's why I want to do them in this order, search, and then if it fails, put it in there.)*

Note that in the last example, the tutor talks in the first-person, but he is actually instructing the student on what to do, so this counts as DPI. The annotation manual (see Appendix C) also provides guidelines for the coders on when an utterance may look like DPI, but it is not. One case worth mentioning is when a tutor creates an incorrect example or follows an incorrect procedure on purpose, to make the student reason about possible incorrect scenarios. In these cases, the "wrong" steps are not marked as instances of DPI.

Direct Declarative Instruction (DDI)   occurs when the tutor imparts knowledge about the domain or a problem. To mark an utterance as DDI, coders must answer the following question in the affirmative: " is the tutor providing a fact that the tutor thinks the student does not already know?" Common sense knowledge is not DDI, and is left unlabeled *(four is more than three)*. Utterances are annotated as DDI if they:

- provide general knowledge about data structures *(so # um by convention the right child is always greater than its parent and the left child is always less than its parent.)*

- convey the results of a given action (*when you did the search already you already found the appropriate place (be)cause that's what you did to find it.*)

- describe the tutoring materials, especially pictures of data structures (*so H one is the header to the first list that tells you where the first node is*).

As for DPI, the annotation manual provides further examples and guidelines. One worth noting is that DPI and DDI are mutually exclusive. Another is that the utterance should not be too far removed from directly addressing the problem; for example *there's a lot of work involved* is not DDI (or DPI, for that matter).

Prompt (PT) occurs when the tutor attempts to elicit a meaningful contribution from the student. To count as a prompt, an utterance may fall into one of seven categories, that can be found in Appendix C. The more common ones are listed here (the utterance is coded only as *Prompt*, but this categorization helps the coder recognize the general *Prompt* label):

- Specific prompt: the tutor is trying to get a specific answer from the student (*so we'll say next of N is equal to +// hmm how do I make it point here?*)

- Fill-in-the-blank: the tutor leaves an utterance incomplete, in the hope the student will complete it. While a fill-in-the-blank prompt is a sub-type of specific prompt, its characteristic close form makes it easier to recognize for the coders; if one day in the future, ChiQat-Tutor were using spoken language, intonation would also play an important part in producing such prompts (*if you want the biggest over here, well, you got to go left to get into this tree and then you go...*).

- Diagnosis: The tutor attempts to determine the student's knowledge state. In general such a prompt is preceded by an incorrect statement or action on the part of the student, to which the tutor reacts by the tutor assessing the situation (*why did you put N over here?*).

Feedback (FB). Up to now, we have discussed annotations for utterances that do not explicitly address what the student has said or done. However, many tutoring moves concern providing feedback to the student. Indeed as already known but not often acted upon in ITS interfaces, tutors do not just point out mistakes but also confirm that the student is making correct steps. While the DAs discussed so far label single utterances, our positive and negative feedback (FB+ and FB-) annotations comprise a sequence of consecutive utterances, that start where the tutor starts providing feedback, or even earlier, when the tutor prompts the student and then provides feedback on their answers (please see our annotation instructions in Appendix C). We opted for a sequence of utterances rather than for labeling one single utterance because we found it very difficult to pick one single utterance as the one providing feedback, when the tutor may include, e.g., an explanation that we consider to be part of feedback. Not only is annotating the start of a feedback episode difficult, so is deciding where it ends; our annotation instructions also provide guidelines in this regard.

Figure 3.3 presents an excerpt of a tutoring session on lists, which includes one example of negative feedback (lines 216–218), and two of positive feedback (219–221, and 226–229, respectively). The labels *START* and *END* when co-occurring with *FB+/-* denote the beginning and end of a positive or negative feedback episode; every utterance within these brackets is part of the corresponding feedback episode.

Positive feedback occurs when the student says or does something correct, either spontaneously or after being prompted by the tutor (as noted above, the prompt is part of the feedback episode - see lines 219 and 226). The tutor acknowledges the correctness of the student's utterance, and possibly elaborates on it with further explanation. Note that utterance 222 (`alright`) by the tutor is not included in the immediately preceding FB+ episode. In this and similar cases, words such as `alright` and `okay` are considered as transition markers in the dialogue, rather than providing feedback.

Negative feedback occurs when the student says or does something incorrect, either spontaneously or after being prompted by the tutor (again, the prompt is part of the feedback episode, see line 216). The tutor reacts to the mistake and possibly provides some form of explanation.

As it happens, all three feedback episodes in Figure 3.3 start with a tutor prompt. This is not necessarily the case; for example, the tutor could react to a student's question or unsolicited utterance, or to an action of the student (recall that the tutor and student are sharing papers on which they are drawing, writing code etc). Indeed, the WOE example in Figure 3.4 includes two examples of positive feedback in response to the student's taking the initiative, i.e., providing content that was not solicited by the tutor (lines 28 and 33 respectively).

Figure 3.3 also illustrates other Dialogue Acts, in particular, a number of DDIs' and DPI's, some of which co-occur with PT's and with start of a feedback episode; line 228 is marked as SI (student initiative) that we will discuss shortly.

Worked-out example (WOE). The annotation of WOEs was superimposed on the other preexisting annotations. Two annotators marked beginning and end of WOEs, similarly to what had been done for feedback.[4]

---

[4]Coders also marked nested WOEs, but since only 21 nested WOEs exist out of 658 total, we will not discuss them further.

Feedback marker	Other DA's	Line & Speaker	Utterance
		215 T:	uh you got to use q@l
[FB- START]	[PT DDI]	216 T:	if it's not the first element, than q@l +...
		217 S:	++ has nothing, q@l is like +...
[FB- END]	[PT DDI]	218 T:	q@l does point to something, right?
[FB+ START]	[PT DDI]	219 T:	q@l isn't +// but if it's the first element, than q@l has never changed, and q@l is still +...
		220 S:	++ null.
[FB+ END]		221 T:	null.
		222 T:	alright.
	[DPI]	223 T:	so if q@l ain't null, then we'll do this.
	[DPI]	224 T:	else, else q@l is null.
		225 T:	and it's the first element.
[FB+ START]	[PT]	226 T:	so what changes?
	[PT DDI]	227 T:	uh there is no q@l, I can't do that sort of a thing, but it used to be a four element list, and now it's going to be a three element list, so, the header, which points to a@l has to point to +...
	[SI]	228 S:	++ the next one.
[FB+ END]		229 T:	the new first element.
	[DDI]	230 T:	so, it's similar to what I have up here.
	[DDI]	231 T:	the one before a@l just leap frogged over a@l.

Figure 3.3 Positive and negative feedback (T = tutor, S = student).

Figure 3.4 shows a WOE in its entirety (26 lines); line 1 is marked with *WOE START*, and line 38, with *WOE END* (not shown in the figure). This example also shows the moves each utterance is labeled with; note that feedback markers are lumped with the other annotations, there is no separate column for feedback as in Figure 3.3. The WOE discusses inserting a new node containing the letter *b* into a list containing the letters *c, k, e, f*. The main pedagogical gist is to show the student how to do so without losing the connection (the pointer) to the node containing *f*.

DA's	Line & Speaker	Utterance
[DDI]	13 T	um so <if we want to insert something such as here >[//] if we want to insert a new node containing g@1, and we want to insert after b@1, we need to be careful how we insert it.
[DDI]	14 T	<because if we were to insert >[//] if we were to break this link first, between b@1 and f@1+...
[PT DDI]	15 T	now nobody knows about f@1, right?
[SI]	16 S	that would just kill it.
[DDI]	17 T	it would just +// you wouldn't have anything that points to f@1, so there's no way for you to find f@1 xxx.
	18 S	okay.
[PT]	19 T	okay?
[DPI]	20 T	so if we were to insert g@1, and we want to insert f@1 to b@1, the first thing that we do is draw it in here.
[DDI]	21 T	this is for the pointer.
[DDI]	22 T	it points to the next node.
[PT]	23 T	okay?
[DPI]	24 T	first thing we would do is we would <check the >[//] find what it's pointing to, which is f@1, and we would set that equal to this.
	25 T	so that g@1 now points to f@1.
	26 S	okay.
[DDI]	27 T	so now b@1 and g@1 both point to f@1.
[FB+ START, SI]	28 S	okay and then you slide g@1 in there b@1 will point to g@1 and g@1 is pointing at f@1.
[FB+ END]	29 T	right.
[DDI]	30 T	because if we were to just get rid of this now nobody is <pointing to f@1 >. [>]
	31 S	[<] <right >.
[DDI]	32 T	and we don't know where f@1 is.
[FB+ START, SI]	33 S	f@1 just gets lost by doing this thing <its still >in line. [>]
	34 T	[<] <right >.
[FB+ END, DPI]	35 T	right, so then we would just come here and delete this.
	36 T	and now we have c@1 k@1 e@1 g@1 f@1.
	37 S	okay.
	38 T	okay so we don't lose f@1.

Figure 3.4 Worked-out example for a linked list problem.

DA's	Line & Speaker	Utterance
[DPI]	533 T	we have these two lists and say we want to switch say # a@l and c@l see each other.
	534 T	ah, we'll keep these in the movie line constant.
	535 T	a@l and c@l see each other and they talk and they find out oh you're in the movie I want to see.
[PT]	536 T	really wanted to see but it was sold out and c@l says the same thing, right?
	537 S	mmhm
[DPI PT]	538 T	so we need to find a way to switch them # ok?
[DDI]	539 T	so what happens is we have one of these guys random guys called temp.
	540 S	right
[DDI]	541 T	comes by and a@l goes to temp and says hey temp can you hold my spot in line for c@l?
	542 T	temp says ok.
[DPI$_j$]	543 T	so we're going to insert temp just like you insert we inserted all the other ones.
[DPI$_j$]	544 T	so temp would come by, we'd go through tell temp to point to m@l.
[DDI PT]	545 T	so he'd point to m@l right #?
	546 T	basically.
[DPI$_j$]	547 T	and we'd tell temp would check the line and see who's pointing to it.

Figure 3.5 Analogy with a movie line for lists.

Analogy.   In our tutoring dialogues, both the tutors use a set repertoire of analogies. For lists, one common analogy is people standing in a line, as shown in Figure 3.5; line 534 is marked with *ANALOGY START* and line 547 with *ANALOGY END*, not shown in the figure (line 533 is included for clarity, and it is actually the start of a long WOE which includes the analogy shown in Figure 3.5 and continues for more than 20 additional lines).

To explain stacks, tutors use analogies with stacks of trays in a cafeteria or with stacks of Lego bricks; one example of the latter is shown in Figure 3.6, where line 2 is marked with *ANALOGY START* and line 11 with *ANALOGY END* (not shown in the figure). As concerns trees, both tutors employ family trees as analogy, see Figure 3.7 (line 145 is marked with *ANALOGY START* and line 148 with *ANALOGY END*).

The annotators were instructed to annotate for the beginning and the end of an analogy episode within a session. For every topic in a session, a tutor usually uses at most one analogy such as to Legos for stacks or to a line for lists. However, the tutor may refer to the analogy several times during a tutoring session. Consequently, in a session, annotators may annotate several episodes for a specific analogy.

DA's	Line & Speaker	Utterance
[PT]	2 T	uh, so if we think of our memory kind of like a, are you familiar with Lego blocks?
	3 S	oh yeah.
[PT DDI]	4 T	so if we think of our our stack, the stack is Legos, every time you want to put on something onto the stack you put a new Lego, right?.
[DDI]	5 T	so first time you put the first Lego we say insert a@1, b@1, c@1, d@1, e@1.
[DPI]	6 T	ok, so we're going to insert a@1 first.
[DDI]	7 T	so +// so the method that we used to put a@1 onto the stack is called push.
[DPI]	8 T	so if I push a@1, I put a@1 onto the stack.
	9 S	right.
[DPI PT]	10 T	ok, and if i push b@1, what I'm doing is adding a new Lego right on top of a@1, ok?
[DPI]	11 T	and if I push c@1, I get another Lego on top of b@1.

Figure 3.6 Analogy with Lego for stacks.

DA's	Line & Speaker	Utterance
[DDI]	145 T	uh a binary search tree is +// one way of looking at it is like a family tree.
[DDI]	146 T	uh each node is what we call a parent.
[DDI]	147 T	each parent has a child.
[DDI]	148 T	and it has a right child and a left child.

Figure 3.7 Analogy with family tree for trees.

What about the student? So far, we have only talked about tutor moves. As we noted earlier, our students talk very little in these dialogues, and often, in response to a tutor's question. However, as has been discussed in the literature, learning often occurs when students are proactive (please see the discussion in Section 2.2 about Chi's active, constructive and interactive learning activities [66, 69], among others). We captured these behaviors by means of annotating for Student Initiative (SI). SI occurs when the student proactively produces a meaningful utterance, by providing unsolicited contributions, or by asking questions. For example, lines 28 and 33 in Figure 3.4 show the student contributing new material that follows from what the tutor says in the previous utterance(s) but was provided spontaneously by the student. Line 228 in Figure 3.3 is also marked as SI, since the student finishes the tutor's utterance and according to the coder, the tutor does not expect the student to complete it. In this specific case, the trailing diacritic at the end of

Table 3.4 Inter-Coder Agreement: Individual Utterances.

Category	Double Coded	Kappa
DPI	10	.7133
DDI	10	.8018
Prompt	8	.9490
SI	14	.8686

tutor utterance 227 would indicate that the tutor was expecting the student to complete the tutor's own utterance, but in this case, the coder judged it otherwise (please see the next section on validating the annotation).

Even if, as we had expected, SIs didn't appear as frequently as tutor moves (see Table 3.8 below for frequencies), as we will see, SI will play a role in some of our regression models, in accordance with our hypothesis that the effectiveness of a dialogue move does not necessarily depend on high frequency of occurrence.

3.2.1.1 Validating the Corpus Annotation

After developing a first version of the coding manual, we refined it iteratively. During each iteration, two human annotators independently annotated several dialogues for one label at a time, compared outcomes, discussed disagreements and fine-tuned the scheme accordingly. This process was repeated until the coding scheme was stable enough that we felt confident we could compute intercoder agreement. To do so, the two annotators independently coded an unseen set of sessions (the *Double Coded* column in Tables 3.4 and 3.5); we then computed inter-coder agreement. For each label, once we reached reasonable agreement, the remainder of the corpus was then independently annotated by the two annotators, each of whom coded half of the remaining sessions. For our final corpus, for the double-coded sessions, we did not come to a consensus label when disagreements arose; rather, we set up a priority order based on topic and coder (e.g., during development of the coding scheme, when coders came to consensus coding, which coder's interpretation was chosen more often), and we chose the annotation by a certain coder based on that order.

There are various approaches to computing inter-coder agreement, but the general consensus in several disciplines – from psychology [77] to content analysis [236, 237], to computational linguistics [57, 103, 24] – is that measures that account for chance agreement need to be used. Hence, we use Kappa [77], which can take values $-1 \leq \kappa \leq 1$. κ is negatively affected by skewed data [103], which happens when a code occurs rarely. This explains why coders had to double code different numbers of sessions for different codes. For example, since Student Initiatives (SI) are not as frequent, we needed to double code more sessions to find a number of SI's high enough to compute a meaningful Kappa (in our whole corpus, there are 1157 SIs but e.g. 4720 Prompts).

Table 3.5 Inter-Coder Agreement: Codes Applying to Episodes.

Category	Double Coded	Kappa
Feedback	5	.6747
WOE	7	.8200
Analogy	5	.5800

The Kappa values we obtained in the final iteration of this process are listed in Tables 3.4 and 3.5. For the labels that apply to several utterances ("episodes") – namely, positive and negative feedback, WOE, and analogy – the κ computation is based on the number of lines that annotators annotate similarly or differently, starting and ending, respectively, at the earliest `Begin` and the latest `End` we found for the label of interest, in the two parallel coded data. For example, consider the FB+ example in Figure 3.3, starting at utterance 219 and ending at utterance 221; suppose that a second annotator had marked the start of the FB+ episode on the next utterance 220 (by the student), and likewise, the end of the FB+ episode on utterance 222, the final tutor utterance. Then, five total utterances would have to be considered, of which only two would represent agreement.

The interpretation of Kappa values is not completely settled. Here, we follow [356] as considering $0.41 \leq \kappa \leq 0.60$ as exhibiting moderate agreement, $0.61 \leq \kappa \leq 0.80$ substantial, and $0.81 \leq \kappa \leq 1$ almost perfect agreement. All the values we obtained indicate from substantial to very high agreement, other than for analogy, where we only found moderate agreement. κ for analogy may have been affected by data sparseness: analogy episodes comprise a very small subset of a session. In fact, analogy is the label for which annotators had to annotate the most sessions while developing the coding scheme. For analogy, annotators first double coded 15 sessions five sessions at a time; every five sessions, they revised the predefined manual. After these 15 sessions, they annotated another five sessions. In contrast, annotators only double coded 5–10 sessions for the other codes before double coding those to compute intercoder agreement. Another source of the disagreements on analogy is due to the difficulty of analogy boundary detection. In fact, annotators are in full agreement on whether analogy is used for a data structure in a session, or not; disagreements are just on the boundaries of analogy episodes.[5]

As a final important note, given our coding scheme, some utterances have more than one label, while others are left unlabeled. Specifically, most student utterances, and some tutor utterances, are not labeled. For example, in Figure 3.3, utterances 215, 222, 225 by the tutor are not labeled; likewise, utterance 220 by the student is unlabeled (student utterances are labeled only when they show initiative, ie., as SIs).

---

[5]Disagreement on boundaries of episodes is a well known problem in discourse analysis [334].

3.3 DISTRIBUTIONAL ANALYSIS

In this section, we report statistics about the elementary dialogue acts, student initiative and episodic strategies that we found in our tutoring corpus.

3.3.1 Elementary Dialogue Acts

Tables 3.6 and 3.7 show distributional statistics about elementary DAs, per topic (recall that tutors were free to skip topics). We report the total number of the specific DAs, the average number per session and per minute. Whereas the difference between the total number of DPI, DDI, Prompt and Feedback between lists and trees on the one hand, and stacks on the other, may appear large, the average number of DAs per minute shows that the difference is due to the dialogues on stacks being shorter, not to fundamental differences in how tutors conduct those interactions. As concerns specific DAs, we can observe that tutors spend about 77% of their time providing instruction (whether procedural or declarative, 18,989 instances total) as opposed to directly involving the student, either by prompting them or reacting to what they do (prompt + feedback, 5682 total instances across the two DAs) As we can also observe, tutors use about eight times more positive feedback than negative feedback; this difference in usage is trending toward significance ($\chi^2 = 5.74, p = 0.057$).

Table 3.6 Elementary DAs Distributional Statistics (Average is reported per session and, in parentheses, per minute)

Topic	Sessions	DPI		DDI		PROMPT	
		Total	Average	Total	Average	Total	Average.
Lists	52	2658	51.1 (3.5)	3926	75.5 (5.2)	1756	33.8 (2.3)
Stacks	46	782	17.0 (3.0)	1871	40.7 (7.0)	638	13.9 (2.4)
Trees	53	3485	65.7 (3.4)	6267	118.2 (6.1)	2326	43.9 (2.3)
Total	151	6925	45.9 (3.4)	12064	79.9 (5.9)	4720	31.2 (2.3)

Table 3 7 Feedback Distributional Statistics (Average is reported per session only, Avg./minute is less than 1, for each cell)

Topic	Sessions	FB+		FB-	
		Total	Average	Total	Average
Lists	52	228	4.4	37	0.7
Stacks	46	97	2.1	15	0.3
Trees	53	560	10.6	55	1.0
Total	151	855	5.7	107	0.7

3.3.2 Student Initiative

As we noted, we did not annotate student turns other than as concerns Student Initiative (SI). SI captures the student's non-prompted contribution to the dialogue. Table 3.8 shows distributional statistics about SI, per session and per topic.

Table 3.8 Student Initiative Distributional Statistics (Average is reported per session only; Avg./minute is less than one, for each cell)

Topic	Sessions	SI	
		Total	Average
Lists	52	354	6.8
Stacks	46	183	4.0
Trees	53	620	11.7
Total	151	1157	7.7

3.3.3 Episodic Strategies

We now turn to the two episodic strategies we annotated for, WOEs and analogies. Tables 3.9 and 3.10 report statistics regarding totals and how many sessions contain a WOE/analogy (columns *Sessions with WOE/Analogy*); number of WOE/analogy episodes per session; average lengths of WOEs / analogies in words and in utterances (with respect to Tables 3.6 through 3.8, we omit frequency per minute, since these are long episodes).

As concerns WOEs, we can observe from Table 3.9 that tutors use many more WOEs for lists and trees than for stacks: the difference between the number of sessions in which WOEs are used for lists and trees on the one hand, and for stacks on the other, is trending toward significance ($\chi^2 = 5.143, p = 0.077$); more frequent WOEs for trees are offset by longer WOEs for lists. Conversely, analogies are used more often for stacks ($\chi^2 = 10.143, p = 0.005$); analogy is used less frequently for lists and the least for trees. However, the analogies used for lists are the longest. As concerns trees being the least likely to be tutored by analogy, we speculate it may be due to the fact that the technical terminology used for trees (parent, daughter, etc) is derived from family trees to start with.

3.4 INSIGHTS FROM THE CORPUS: PEDAGOGICAL MOVES AND LEARNING

Our interest in collecting human–human tutoring dialogues is two fold: theoretically, we want to gain insight into which pedagogical strategies are more effective for learning; practically, the most effective strategies will be the ones we will include in the ChiQat-Tutor ITS. The question is, how to find

Table 3.9 Worked Out Examples (WOEs) Distributional Statistics (Utts. stands for Utterances)

Topic	Sessions	Sessions with WOEs	Total WOEs	WOEs per Session	Words per WOE	Utts. per WOE
Lists	52	50	180	3.5	498.3	48.3
Stacks	46	23	24	0.5	615.5	68.5
Trees	53	51	454	8.6	212.5	24.0

Table 3.10 Analogy Distributional Statistics (Utts. stands for Utterances)

Topic	Sessions	Sessions with Anal.	Total Anal.	Analogies per Session	Words per Anal.	Utts. per Anal.
Lists	52	21	54	1.0	334.2	31.9
Stacks	46	40	63	1.4	121.2	10.8
Trees	53	16	22	0.4	54.5	5.5

evidence for the effectiveness of those strategies. Early on, we adopted multiple regressions as our preferred way of inquiry: we correlate the frequencies of the tutoring behaviors in a session (the predictor variables) with the amount of learning in that session (the predicted variable). The beta weights, the partial regression coefficients, indicate the relative strength of the relations between the relevant pedagogical moves and strategies, and the learning outcomes. This method is appropriate because it does not assume that the more effective tutoring moves are necessarily more frequent, in absolute numbers, than the less effective ones; rather, this method uncovers whether variation in the frequency of one type of pedagogical move is more strongly related to variation in learning outcomes than another type of move, regardless of each move's relative frequency (please see [329] for further comments on this topic).

Because of the complexity of human dialogues, we conducted our experiments in several rounds.

1. We started by exploring models that only include dialog acts that apply to individual utterances;

2. Then, we developed models that include sequences of dialog acts;

3. Finally, we explored models that include episodic strategies, namely, WOE and analogy.

Models of types 2 and 3 both include what could be considered "groups" of dialog acts. The difference is that models of type 2 involve sequences of two or three consecutive dialog acts, where these sequences are created by exhaustively considering all possible combinations of DAs; most of these sequences do not have any a priori cognitive import. On the other hand, in models of type 3, we focus on longer sequences of dialogue acts that constitute an episode such as a WOE or an analogy; such episodes are considered as a unit which can have a potential impact on learning. In this respect, *Feedback* almost plays a dual role, as an individual dialog act and as an episode. As we noted, *Feedback* is coded as an episode itself, with a begin and an end, like WOE and analogy. However, we will enter it in the regression as an individual dialogue act, because in general *Feedback* episodes are short, comprising 26 words and 3.5 utterances on average. This is one order of magnitude shorter than WOE and analogy episodes, please see Tables 3.9 and 3.10 (analogies for stacks and trees are not as long, but still double the length of feedback episodes). Finally, all our models include pre-test as a co-variate; pre-test will in fact be retained in all significant models, will negatively correlate with learning and will explain a relatively large part of the variance. This reflects well-known educational findings that prior knowledge affects learning, and specifically, the amount of learning that can potentially take place [186, 107]. Another well-known variable that affects learning is time-on-task, that we will approximate with session length. The results on session length will be more nuanced, as we will illustrate below.

3.4.1 Individual Dialog Acts (Type 1 Models)

These models investigate the correlations between the presence of individual dialog acts and learning gains: for each topic in a session (i.e., for one individual student), the predicted variable is learning gain; the predictor variables are listed in Table 3.11.

Table 3.11 Predictor Variables used in Multiple Linear Regression Models

Feature	Description
Pre-test	Pre-test score
Length	Length of the tutoring session
FB+/-	Number of positive/negative feedback
DPI	Number of direct procedural instructions
DDI	Number of direct declaration instructions
PT	Number of prompts
SI	Number of student initiatives

Table 3.12 shows the results. The column M refers to different models we ran. Model A only includes pre-test, Model B adds session length to pre-test and Model C adds to pre-test all the DAs. Note that Model C does not include length of session. We did run all the equivalent models to Model C's including length. The R^2's stay the same (literally, to the second decimal digit), or minimally decrease. However, in all these Model C's that include length, no DA is significant, partly because we found that certain DAs, like

Table 3.12 Regression Models (Type 1)

Topic	M	Predictor	β	R^2	P
Lists	A	Pre-test	−.47	.202	< .001
	B	Pre-test	−.43	.290	< .001
		Length	.01		< .001
	C	Pre-test	−.500		< .001
		FB+	.020		< .01
		FB-	.039		ns
		DPI	.004	.377	< .1
		DDI	.001		ns
		SI	.005		ns
		PT	.001		ns
Stacks	A	Pre-test	−.46	.296	< .001
	B	Pre-test	−.46	.280	< .001
		Length	−.002		ns
	C	Pre-test	−.465		< .001
		FB+	−.017		< .01
		FB-	−.045		ns
		DPI	.007	.275	ns
		DDI	.001		ns
		SI	.008		ns
		PT	−.005		ns
Trees	A	Pre-test	−.739	.676	< .001
	B	Pre-test	−.733	.670	< .001
		Length	.001		ns
	C	Pre-test	−.712		< .001
		FB+	−.002		ns
		FB-	−.018		ns
		DPI	−.001	.667	ns
		DDI	−.001		ns
		SI	−.001		ns
		PT	−.001		ns
All	A	Pre-test	−.505	.305	< .001
	B	Pre-test	−.528	.338	< .001
		Length	.06		< .001
	C	Pre-test	−.573		< .001
		FB+	.009		< .001
		FB-	−.024		ns
		DPI	.001	.382	ns
		DDI	.001		ns
		SI	.001		ns
		PT	.001		ns

DPI, have a very high degree of collinearity with session length, making those two variables (DPI and length) almost interchangeable. We consider models that include length as a significant variable as less explanatory than the Model

C's in Table 3.12, which do not include length: finding that a longer dialogue positively affects learning does not tell us what happens during that dialogue which is conducive to learning.

As evidenced by the table, only FB+ and DPI provide significant or marginally significant contributions (in the table, we also include the non-significant DAs to show their β coefficients; we will not do so any longer moving forward). The fact that FB+ is positively correlated to learning in at least some of our models, points to the power of the correlational approach, since FB+ occurs about 1/14 of the times that DDI occurs and 1/5 of the times that Prompt occurs; but these other two DAs are never significant in any models. However, R^2 improves only for lists, and for the models that don't distinguish among topics (indicated by *All*); for stacks and trees, R^2 minimally decreases from model A and/or B when adding the DAs. Also note that the β weights on the pre-test are always negative in every model, namely, students with higher pre-test scores learn less than students with lower pre-test scores. This is an example of the well-known *ceiling effect*: students with more previous knowledge have less *learning opportunity*. Also noticeable is that the R^2 for the trees models is much higher than for lists and stacks, and that for trees, no DA is significant in this sort of models (as we will see, there will be significant models of type 2 and 3 that include sequences of DAs for trees).

3.4.2 Sequences of Dialogue Acts (Type 2 Models)

The models in Table 3.12 are our most basic models, as they explore correlations between individual DAs and learning (always keeping pre-test score as a co-variate). However, as research in various areas of linguistics has abundantly shown, DAs do not occur in isolation but perform specific functions given the context. From the point of view of ITS design, models that include sequences of DAs are bound be more useful than those that include individual DAs out of context. For example, take model C for lists in Table 3.12. It tells us that positive feedback (FB+) and Direct Procedural Instruction (DPI) positively correlate with learning gains. However, first, this obviously cannot mean that our ITS should only produce FB+ and DPI. The ITS is interacting with the student, and it needs to tune its strategies according to what happens in the interaction. Second, and more relevant for the current discussion, Model C does not even tell us if FB+ and DPI should be used together or independently; and if together, which of the two should precede the other.

Hence, in the remainder of this analysis, we will investigate DAs in context. The first and simpler way of doing so is to analyze sequences of DAs and their correlations with learning. This of course begs the question, which sequences of DAs, how long they should be, etc. To answer this question, we follow established practice in the NLP literature [217] to build *n-grams*, i.e. sequences of n "items", where the "items" are all the different values that the label of

DA	Line & Speaker	Utterance
[DDI]	327 T:	here's where t@1 was and here's where s@1 is going.
	328 S:	so s@1 equals t@1.
[FB+ END]	329 T:	yeah
[PT]	330 T:	and does it make any difference what order I do that in?
	331 S:	probably.
[PT]	332 T:	before I update t@1 or after I update t@1?

Figure 3.8 DA sequences: [DDI, FB+] and [FB+, PT] bigrams; [DDI, FB+, PT] trigram

interest takes: in our case, the label is DA, and its values are DPI, DDI, etc. As is well known, as soon as $n > 3$, the number of actual sequences occurring in a language corpus decreases dramatically, namely, the data becomes very sparse [6] in what follows, we will investigate sequences of 2 (bigrams) or 3 DAs (trigrams). Innovatively, we will also permit gaps in the sequence, up to a small g ($0 \leq g \leq 3$). This will allow us to focus on the DAs of interest in the sequence, even if another utterance intervenes.

For example, consider the excerpt in Figure 3.8. When we generate bigrams and trigrams, we follow these steps:

1. We generate any sequence of 2 or 3 utterances from the data. For the excerpt in the figure, the set of bigrams is: [327,328]; [328,329]; [329,330]; [330,331]; [331, 332]; the set of trigrams is: [327,328,329]; [328,329,330]; [329,330,331]; [330,331,332].

2. We first build the bigrams and trigrams of DAs, where each consecutive utterance is labeled by a DA. For this example, this results in one single bigram [FB+,PT], and no trigrams (because of the unlabeled utterances 328 and 331).

3. We now relax the constraint that there cannot be any gaps g in the bigram and trigram sequences. $g = 1$ results in bigrams [DDI,*,FB+], [*,FB+,PT] and [PT,*, PT]; and trigrams [DDI,*,FB,PT], and [FB,PT,*,PT], where * matches any other DA that may occur in that position. $g = 2$ results in bigrams such as [DDI,*,*,PT] where we allow two intervening DAs between DDI and FB+, but also [*,DDI,*,FB+], which covers a sequence consisting of any DA, followed by DDI, followed by any DA, followed by FB+; and [DDI,*,*,PT], i.e.,

---

[6] It is hard to obtain meaningful frequencies for n-grams when $n > 5$ even from very large text collections such as the 14 billion word iWeb corpus from Brigham Young University, or even the 155 billion Google Books n-gram corpus of American English (https://www.english-corpora.org/).

DDI followed by any other two DAs, followed by PT. $g = 3$ results in even more potential sequences such as [DDI,*,*,PT,*] that we have no room to include.[7]

The astute reader will remember that some of our utterances are marked with more than one DA, see for example the sequence 15-18 in Figure 3.4, in which utterance 15 is labeled as both PT and DDI. Then, any bigram and trigram including 15, and for all possible lengths of gaps[8], would have to be duplicated, to include once PT and once DDI. For example, a bigram/trigram starting at 15, with no gaps, would appear once as [PT,SI]/[PT,SI,DDI]; and once as [DDI,SI]/[DDI,SI,DDI].

Once we generate all bigrams and trigrams with gaps, as just described, we investigate which DA sequences may affect student learning. We ran thousands of linear regression experiments to compare the results with those in Table 3.12: models including a single specific bi-gram $\mathcal{B}$ or trigram $\mathcal{T}$ (with or without gaps); that same bigram $\mathcal{B}$ or trigram $\mathcal{T}$ with pre-score as a co-variate; finally, $\mathcal{B}$ or $\mathcal{T}$ with pre-score and session length as co-variates.

For each model just described:

1. We index the corpus according to the length of the n-gram, 2 or 3 (bigram or trigram).

2. We generate all the permutations of all the DAs we annotated for within the specified length; we count the number of occurrences of each permutation in the corpus by topic, as appropriate. Namely, we count how many occurrences of [DDI,FB,PT] occur in the corpus, separately, for lists, stacks and trees; but we also count occurrences of say [FB,FB,FB], or [SI,SI,SI], even if presumably they do not appear. As mentioned, these sequences also include gaps g of a specified length, with $0 \leq g \leq 3$ (please see [63] for further details). Gaps can be discontinuous, namely, $g = 2$ may be instantiated as [*,DDI,*,FB], i.e., a bigram [DDI,FB] preceded by any other DA, and with one intervening DA.

3. We run linear regressions on the models listed above, generating actual models by instantiating a generic DA bi- or trigram with each DA sequence we generated in step 2, whose frequency is different from zero.

4. We collect only those models which satisfy the following condition: the whole model and every predictor must be at least marginally significant ($p < 0.1$).

---

[7]There is a subtle difference between sequences that include a * at beginning or end, and the same sequence without that *: e.g. [*,DDI,FB] covers more bigrams that simply [DDI,FB].

[8]The bigrams and trigrams with gaps were generated after indexing the corpus by means of Lucene Apache, a text search engine library https://lucene.apache.org/core/.

3.4.2.1 Bigram Models

Whereas we did find significant models that include bigrams, their R^2 is not higher than the models presented in Table 3.12. However, we will briefly discuss them since they in fact highlight the significance of the two dialog acts, DPI and FB, that already resulted in significant models when taken individually, as shown earlier in Table 3.12.

Table 3.13 shows the role played by the bigram [DPI, FB] for lists, and to a minor extent, for stacks. Specifically, the sequence that includes DPI followed by any type of feedback (FB, FB+, FB-) produces significant models when the model includes pre-test and/or length. Different models are derived if we include the type of feedback (FB+, FB-), or not (FB), as in the second list model in Table 3.13. Note these are not necessarily "straight" n-grams, namely they may include gaps. If the number in the column *Gap Length* is greater than zero, it means that the model was obtained with that gap length. For example, the first model for lists in Table 3.13 was obtained with a gap length = 2, meaning, the frequency of the bigram [DPI,FB-] is the sum of the frequencies of all combinations of this bigram instantiated with a 2-utterance gap, such as [DPI,FB-,*,*],[DPI,*,FB-,*],[DPI,*,*,FB-], etc (a total of 6 such sequences).

Table 3.13 [DPI, Feedback] Models

Topic	Predictor	β	R^2	P	Gap Length
Lists	Pre-test	−.513	.235	< .001	2
	[DPI, FB-]	.039		< .001	
	Pre-test	−.489		< .001	
	Length	.011	.333	< .05	1
	[DPI, FB]	.016		< .05	
	Pre-test	−.492		< .001	
	Length	.011	.339	< .05	1
	[DPI, FB+]	.019		< .001	
Stacks	Pre-test	−.401	.342	< .001	2
	[DPI, FB-]	−.187		< .1	

For lists, these models are not as predictive as Model C in Table 3.12; however, they may be more useful from the point of view of tutorial strategies to include in an ITS, since they provide more information than simply noticing that DPI and FB+ individually are correlated with learning. As we noted earlier in this section, type C models in Table 3.12, do not even tell us if FB+ precedes or follows DPI, but the bigram [DPI,FB+] gives us the answer, FB+ follows. As for the pedagogical reasons behind this, the [DPI, FB+] bigram with gap=1 tells us that when the tutor provides direct instruction on solving a problem (DPI), and very shortly afterward the tutor provides some positive feedback (FB+), this correlates with increased learning: we speculate this may happen because the student will have tried to apply that DPI (these regression models do not include what the student does, other than when they take initiative (SI) because we did not have an efficient way of coding for student's actions. When we discuss learning with the ChiQat-Tutor system, logs of student actions will provide us with a more thorough picture of the learning process).

For stacks, the lone significant model we obtained which includes a DA bigram is actually more predictive than Model C for stacks in Table 3.12 (that model includes FB+ but not DPI). Interestingly, the β weight is negative for the sequence [DPI, FB-] in the stacks model. We speculate that stacks are so easy for students that the presence of negative feedback in relation to procedural instruction highlights that the student knowledge deficiency is very high; this finding does not directly translate into guidelines for ITS design however. No models including the bigram [DPI, FB] with gaps of any length are significant for trees.

A few other bigram models are also significant, and interestingly, include FB in some fashion. However, since they are not more explanatory than the models in Table 3.12, we will not discuss them further, please see [63] for further details.

3.4.2.2 Trigram Models

No bigram model was found to have a better fit that the models in Table 3.12, which include individual and independent bigrams, even if the significant bigrams provided interesting insights as concerns the order of DAs. Those same DAs (FB and DPI) appear in the trigram models that achieve slightly higher R^2 than the models in Table 3.12; interestingly, these trigrams include PT (Prompt) or SI (Student Initiative) namely, DAs that explicitly invite the partner to contribute to the conversation (PT), or that indicate such a spontaneous contribution to the conversation. These models can be found in Table 3.14 (the two trees models differ in whether FB is generic, or FB+, i.e., positive). The improvements in R^2 over the models in Table 3.12 are slight; the largest is 0.12 for stacks. It is interesting to note that for stacks, as for bigrams, the β coefficient between the DA trigram and post-test is still negative.

Table 3.14 Trigram Models

Topic	Predictor	β	R^2	P	Gap
Lists	Pre-test	−.463		< .001	
	Length	.011	.415	< .05	0
	[PT,DPI,FB]	.266		< .01	
Stacks	Pre-test	−.52	.416	< .001	1
	[DDI,FB,PT]	−.06		< .01	
Trees	Pre-test	−.746	.732	< .001	1
	[FB+,SI,DDI]	.049		< .01	
	Pre-test	−.746	.732	< .001	1
	[FB,SI,DDI]	.049		< .01	

The model for lists adds *PT* at the beginning of a bigram that had already been found to contribute to a significant model. For stacks DDI (declarative instruction) substitutes DPI (procedural instruction), but otherwise, the significant trigram is similar to that for list. For trees, the FB/DDI order reverses (FB comes before DDI), and SI (Student Initiative) is inserted in the middle. [FB,SI, DDI] denotes that, after the tutor provides feedback, the student takes the initiative, and the tutor provides some information (DDI) the student probably did not know, or that the tutor deems the student needs to be reminded of. It is not surprising that Prompts and SIs play an important role, even if they are significant only in association with certain tutor moves, and not individually; PT and SI are never significant in the models in Table 3.12. This is similar to the findings in [266] where a student's novel answer (that could be considered a type of student initiative) was less strongly correlated with learning in isolation than in the context of a tutor move either preceding, i.e., a tutor's Short Answer Question, or following, e.g., a tutor's recap.

3.4.3 Episodic Strategies (Type 3 Models)

We now turn to regression models of type 3, which include cognitively motivated sequences of utterances that can be considered as units, namely, WOEs and analogies.

Table 3.15 **Description of WOE Features**

Type	Feature	Description
Descriptives	WOE	Number of WOEs
	WOE-Words	Length of in words
	WOE-Utts	Length of in utterances
Within WOE episodes	WOE-DPI/DDI	Count of DPI/DDI
	WOE-PT	Count of prompts
	WOE-FB/FB+/FB-	Count of feedback (neutral / positive / negative)
	WOE-SI	Count of student initiatives
Outside WOE episodes	NoWOE-DPI/DDI	Count of DPI/DDI
	NoWOE-PT	Count of prompts
	NoWOE-FB/FB+/FB-	Count of feedback (neutral / positive / negative)
	NoWOE-SI	Count of student initiatives

3.4.3.1 Worked-Out Examples

All potential features related to WOEs that we used in our experiments are presented in Table 3.15. The first regression analysis we ran that includes WOEs, simply adds the number of WOEs per session, which does not correlate with learning gains (actually it does for stacks, but the correlation is negative). Next, we explore models where we differentiate between DAs that occur within a WOE episode (*WOE-DA*) and those that occur outside of WOE episodes (*NoWOE-DA*). We systematically ran every regression analysis that includes: pre-test, length of session and all the features listed in Table 3.15. This results in models with a better fit (see Table 3.16) but not for trees (the tree model in Table 3.16 is of a different nature and we will discuss it shortly).

Table 3.16 WOE: Multiple Regression Models

Topic	Predictor	β	R^2	P
Lists	Pre-test	−0.442		< .01
	WOE-PT	−.0006	.485	< .1
	WOE-Utts	.002		< .1
	NoWOE-FB+	.005		< .1
Stacks	Pre-test	−.37	.606	< .005
	WOE-FB+	0.077		< .005
	WOE-PT	−.021		< .05
Trees	Pre-test	−.736	.737	< .0001
	WOE-[FB+,SI,DDI]	.037		< .005

From Table 3.16, it is apparent that not many WOE features may generalize across topics; for example, the length of WOEs in utterances (*WOE-utts*) is marginally positively correlated with learning gain only for lists, namely, it is only for lists that longer WOEs positively affect learning. The feature that is present in the models for all three topics is positive feedback (FB+), but it appears in different guises in each. For stacks, FB+ within WOEs (*WOE-FB+*) is effective, and for trees, it correlates with learning as part of the sequence [FB+,SI,DDI]. For lists, it is FB+ outside of WOEs (*NoWOE-FB+*) that marginally correlates with learning. One surprising observation is that for lists and stacks, prompts within WOEs (*WOE-PT*) are negatively correlated with learning gains (even if the β coefficients are low, and for lists, it is only a marginally significant correlation). This contrasts with our previous findings on the benefit of prompts for learning when included in DA sequences (see discussion in Section 3.4.2.2). The result for prompts included in WOEs seems to suggest that when the tutor is demonstrating a solution, as in a WOE, the tutor should complete the demonstration before inviting students to participate. A related finding is that, on average, tutors used more prompts within WOEs for stacks (11.1), than for lists (7.7), than for trees (3.3). As we have remarked elsewhere, stacks are considered the easiest among these three data structures, and this may be the reason tutors include more prompts in WOEs for stacks; but apparently, this strategy may backfire. Indeed as we noted earlier, the mere number of WOEs negatively correlates with learning for stacks.

Finally, a comment on trees: none of the WOE features in Table 3.15 resulted in better models for trees. In this case, we took the best previous model for trees which includes pre-test and the DA trigram [FB,SI,DDI] (see Table 3.14). A model that, in addition to pre-test, only includes the

Table 3.17 Analogy Features used in Multiple Linear Regression Models

Type	Feature	Description
Descriptives	AN	1 if analogy used, otherwise 0
	AN-Words	Length in words
Within analogy episodes	AN-DPI/DDI	Number of DPI/DDI
	AN-FB/FB+/FB-	Number of feedback (neutral / positive / negative)
	AN-PT	Number of prompts
	AN-SI	Number of student initiatives
Outside analogy episodes	NoAN-DPI/DDI	Number of DPI/DDI
	NoAN-FB/FB+/FB-	Number of feedback (neutral / positive / negative)
	NoAN-PT	Number of prompts
	NoAN-SI	Number of student initiatives

occurrences of this trigram of DAs within WOEs (slightly modified with FB+ as opposed to neutral FB), results in a slightly improved $R^2 = .737$ for trees as well.

3.4.3.2 Analogies

We ran linear regressions that include features of analogy episodes, similar in nature to the features we used for WOEs, including differentiating between DAs within analogies (*AN-DA*) and outside of analogy episodes (*NoAN-DA*). These features are listed in Table 3.17.

Table 3.18 includes the statistically significant models we obtained by running all possible combinations of features, including analogy features. We distinguish between Models A and B, not to be confused with Models A, B and C of Type 1 (see Table 3.12). Features included in Model A are: pre-test, length of the session, individual dialogue acts (not differentiated as concerns within and outside of analogy episodes) and presence of analogy (AN). AN correlates with learning (whether significantly or marginally) only for lists. Interestingly, dialogue acts similar to those we have already observed in Type 1 and Type 2 models contribute in Model A (FB+ for lists, DPI and PT for stacks). However, this does not hold for trees, where no feature is even marginally correlated with learning, other than pre-test (note that Model A for trees here has a slightly worse fit than Model A for trees in Table 3.12. In the current model A, pre-test was left as the only significant features, but the regression was run with the additional features mentioned above; instead Model A for trees in Table 3.12 was run only with pre-test to start with.)

Model B removes undifferentiated DAs, and adds analogy-based features beyond mere presence of analogy; all models B result in a better fit (improved adjusted R^2). However, surprisingly, most DAs within analogy episodes

Table 3.18   Analogy: Multiple Linear Regression Models

Topic	Model	Predictor	β	R^2	p
List	A	Pre-test	-0.330		$< .001$
		FB+	0.015	0.388	$< .1$
		DPI	0.005		$< .1$
		AN	0.145		$< .1$
	B	Pre-test	-0.42		$< .005$
		NoAN-DPI	0.0054		$< .05$
		AN-SI	-0.062		$< .1$
		AN	0.308	0.458	$< .005$
		AN-PT	-0.021		$< .05$
Stack	A	Pre-test	-0.408		$< .001$
		DPI	-0.01	0.348	$< .1$
		PT	0.007		$< .1$
	B	Pre-test	-0.418		$< .001$
		NoAN-PT	-0.011	0.414	$< .1$
Tree	A	Pre-test	-0.755	0.696	$< .001$
	B	Pre-test	-0.756		$< .001$
		AN-DDI	0.026	0.725	$< .1$
		AN-PT	-0.109		$< .1$

negatively correlate with learning, with only *AN-DDI* marginally positively correlating with learning, in Model B for trees. In lists, the presence of analogy per se correlates with learning; however, both the prompts (*AN-PT*) and student initiative (*AN-SI*) inside analogy negatively correlate with learning (even if the trend is only marginally significant). For trees, *AN-PT* also negatively correlates with learning. These results while surprising follow in the same mold as the negative correlation between prompts within WOEs and learning we found for stacks, and we discussed in the previous section. In this case for lists and trees, while analogies may be useful in general (at least for lists), trying to engage the student with prompts within an analogy is not helping the student learn; conversely but for similar reasons, a student taking initiative within analogies for lists is also counterproductive.

3.5   SUMMARY: INSIGHTS FROM HUMAN TUTORING ANALYSIS

In the previous section, we have explored the correlation between features of our dialogues, and learning. We adopted a multiple regression analysis, and we explored three types of models: Model 1, which includes individual dialogue acts; Model 2, which includes sequences of DAs of length 2 (bigrams) and 3 (trigrams) and Model 3, which adds meaningful episodes in the dialogues, namely, WOEs and analogies. Table 3.19 summarizes the results, reporting for every topic, the most explanatory model for each type of model – actually for Model 3, we report separate results for WOE and analogy. The features column lists all the features different from pre-test that are included in the corresponding model, whether they are statistically or marginally significant. If a feature is followed by "(-)", it means the correlation is negative, otherwise the correlation is positive.

Table 3.19 highlights a few of the findings from our analysis. First, it is clear that adding groups of DAs, whether as sequences (Model 2), or as meaningful episodes (Model 3), as opposed to individual DAs (Model 1), improves the fit of the model. R^2 increases for all topics, from 7% for trees, to 11% for lists, to 33% for stacks. It is also interesting to note that the largest R^2 is always obtained when adding WOE features to the regression, even if the difference in R^2 with respect to Model 2 and/or to Model 3-Analogy, is often rather small.

We will now analyze which features appear in the models in Table 3.19. First, recall that pre-test is included in all of these models, that it is always negatively correlated to learning, and that in fact it is the feature that always has the largest β weight. Across all features, feedback (FB), and in particular, positive feedback (FB+), appears in one guise or the other in all models, for all topics, excluding the Analogy models; the correlation is almost always positive, excluding Models 1 and 2 for stack. Other DAs such as DDI and DPI also appear in some models but not as pervasively as FB. More specifically, DPI appears in models for Lists, whereas DDI appears in models for Stacks and Trees. Finally, Prompts (PT) offer an interesting quandary: other than when PT appears in the list trigram model, the correlation between PT and learning is negative. This is puzzling since PT is a DA whose aim is to engage the student, and student engagement has been shown to be conducive to learning. It is of note though, that almost all negative correlations between PT and learning occur when PT appears within WOE or Analogy (cf. WOE and Analogy models for both lists and stacks). It is possible that inviting students to contribute within this longer episodes turns out to be counterpro-

Table 3.19 Summary of Model Correlations

Topic	R^2	Model Type	Features (besides Pre-test)
List	.377	Model 1	FB+, DPI
	.415	Model 2: Trigram	Length, [PT,DPI,FB]
	.458	Model 3: Analogy	NoAN-DPI, AN, AN-SI (-),AN-PT (-)
	.485	Model 3: WOE	WOE-PT (-), WOE-Utts,NoWOE-FB+
Stack	.275	Model 1	FB+ (-)
	.414	Model 3: Analogy	NoAN-PT (-)
	.416	Model 2: Trigram	[DDI,FB,PT] (-)
	.606	Model 3: WOE	WOE-FB+, WOE-PT (-)
Tree	.667	Model 1	(None)
	.725	Model 3: Analogy	AN-DDI, AN-PT (-)
	.732	Model 2: Trigram	[FB+,SI,DDI]
	.737	Model 3: WOE	WOE-[FB+,SI,DDI]

ductive, perhaps adding to the considerable cognitive load students undergo when engaging with a WOE or an analogy.

Student Initiative (SI) rarely appears in our models, with a negative correlation in the analogy model for lists, and in the trigram model for trees. As we just noted, it may be the case that a more interactive execution of an analogy which involves prompts and student initiative adds to the student's cognitive load; and earlier in Section 3.4.2, we observed how an SI may be more effective only in a larger context, as our trigram model points to.

A word now on models for Stack (across model types): the vast majority of correlations are negative, namely, the fewer occurrences of X, where X varies across several features, in a stack session, the more the student learns. This seems to point to the fact that what's of the essence for stacks is brevity (even if length did not correlate with learning, whether negatively or positively, in Models of Type 1 for stacks). As we pointed out earlier, this may be related to the fact that stacks are easy as a data structure; hence, even students who do not know stacks should not require too much tutoring to master them.

In conclusion, in this chapter, we have illustrated the methodology we adopted to analyze our human-human tutoring dialogues for features that correlate with student learning. We discussed how we annotated the data, and how we ran many regressions involving different dialogue moves, individually or in episodes, to uncover correlations with learning. In the next few chapters, we will illustrate how these findings informed the development of the ChiQat-Tutor ITS: specifically, in Chapter 4, we will discuss how tutoring moves and sequences thereof were embodied in ChiQat-Tutor's pedagogical strategies; in Chapter 5, we will discuss our extensive experiments in which students learned linked lists with versions of ChiQat-Tutor that differ in the kind of tutoring strategies they use.

ChiQat-Tutor and its Architecture

With Omar AlZoubi

Jordan University of Science and Technology

With Christopher Brown

United States Naval Academy

CONTENTS

A T A CONCEPTUAL LEVEL, Intelligent Tutoring Systems include four components, as we discussed earlier in Section 2.4:

> the domain model which stores domain knowledge, the student model which stores the current state of an individual student in order to choose a suitable new problem for the student and the tutor model which stores pedagogical knowledge and makes decisions about when and how to intervene. [10, p. 5]

Additionally, the classic ITS architecture also includes a *user interface* module that allows the student to interact with the system [392, p. ii].

Given this abstract architecture, in reality, there is great variability as concerns how it is realized in a specific ITS, and how each component is actually implemented. In some cases, a conceptual module does not exist per se other than as dynamically being "embodied" by the corresponding functionality when the ITS is in use; this happens especially for the student model and the interface.

In this chapter, we will describe ChiQat Tutor at that same conceptual level, but pointing out when the mapping is more tenuous; for example, ChiQat-Tutor does not include an explicit student model, but the functionality of the student model is mainly achieved when computing which type of feedback to provide to the student. For ease of exposition, and as it reflects how the conceptualization and development of ChiQat-Tutor proceeded, we call the version of ChiQat-Tutor that included feedback but no Worked-Out Examples, ChiQat-Tutor-v1; whereas the version that includes both feedback, WOEs and analogy, is called, ChiQat-Tutor-v2.[1]

In this chapter, we focus on a conceptual overview of the full system and of the relations between components, as presented in Figure 4.1. We will not cover the description of the implementation, which can be found in Chapter 7. Chapter 7 also provides a road map on how to use our modular architecture to extend the coverage of the ITS.

4.1 THE DOMAIN MODEL

ChiQat-Tutor's domain model contains four main components: problem definitions; constraints representing successful solutions; WOEs definitions and the Procedural Knowledge Model, which represents a solution space for the problems in the domain.

4.1.1 Problem Definitions

A problem is given to the student in the form of a textual description and one or more initial scenarios. A problem definition also includes an exemplary correct solution, which will be used by the solution evaluator to check the solution provided by the student.

ChiQat-Tutor supports two types of problems. The first kind of problems can be solved interactively, step-by-step, as shown in Figure 4.2.

Students can enter a command into the system (command line at bottom of right window, greyed out in Figure 4.2) and the system simulates the effect of that command, immediately showing the effect of the action on the simulated scenario (left bottom window); in this particular case, the student has created

---

[1]In our early papers [129, 130, 132, 131], ChiQat-Tutor-v1 was called iList.

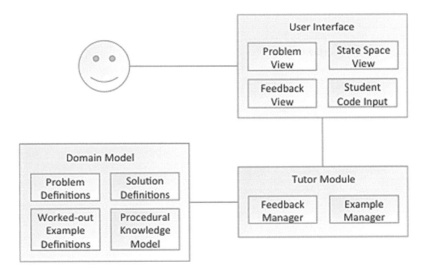

Figure 4.1 ChiQat-Tutor High Level Architecture.

a new node N which is shown in the bottom window as being undefined, as represented by the *?* in its data component (we will discuss the interface to ChiQat-Tutor in detail in Section 4.2)

The second type of problems requires writing a complete snippet of code, typically involving iterative constructs like loops, see Figure 4.3.

Problems of this type usually introduce more than one initial scenario and ask the student to write code that should work correctly in all of them. This encourages the student to abstract away the specific details of a scenario and think about more general algorithms for solving problems on a wider range of situations.

The curriculum for linked lists in ChiQat-Tutor is currently composed of seven problems, five of the first type and two of the second type (linked lists is the most developed topic in ChiQat-Tutor so far; in Section 6.1, we will describe the BST problem set). These problems have been carefully crafted based on our experience as computer science educators and on published CS curricula, such as ACM [1]. The goal is to challenge the students with common difficulties in manipulating linked lists. The problems are defined in ChiQat-Tutor using a human-readable XML format, which makes it easy to add new problems as needed. Chapter 7 will provide full details on how problems are authored.

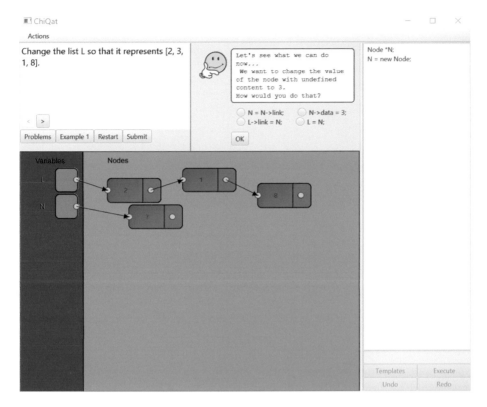

Figure 4.2 Screenshot of ChiQat-Tutor – Step-by-step problem.

4.1.2 Solution Definitions

For this component of the domain model we adopted Constraint-Based Modeling, that we discussed in Section 2.4.

In the linked list domain, a correct solution should exhibit several properties. For example, a list should contain the correct values, as specified in the description of each problem; lists should be free of cycles; lists should not terminate with undefined or incorrect pointers; no nodes should be made unreachable from any of the variables, i.e., lost in the heap space; nodes should be correctly deleted when necessary (this applies specifically to non-garbage collected languages, like C++). With these properties in the form of constraints, ChiQat-Tutor can catch many common mistakes students make.

4.1.3 Worked-Out Examples

In Sections 2.4.2 and 3.4.3, we discussed Worked-Out Examples (WOEs) and analogies both in general, and as pedagogical strategies that we analyzed in our human–human tutoring dialogues. In its second incarnation,

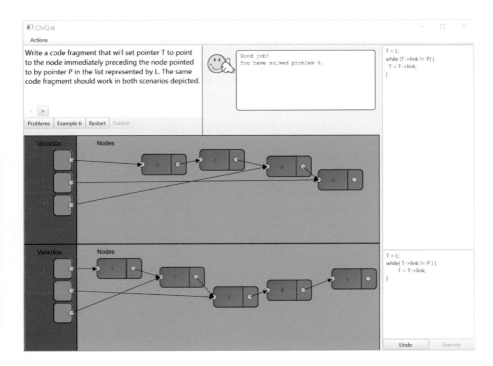

Figure 4.3 Screenshot of ChiQat-Tutor – Block of code problem

ChiQat-Tutor-v2 includes WOEs and analogies. At its core, a WOE is described via a definition (in XML) which is based on a cyclic directed graph, where each node is considered a WOE step. Each step is related to an action, such as the virtual tutor giving some instruction or an instruction being added to a list of solution steps. The steps are then connected to other steps. All our WOEs include a series of steps leading from problem to solution in a serial fashion. Whereas the definitions in principles allow for branching, in practice, we did not make use of this capability in ChiQat-Tutor.

As we will see later, we have two types of analogies: one will simply be an initial pop-up analogy that will be shown to the student before starting their work and the second will be defined as a WOE, but the steps in the example will use entities from the analogical domain, for example, people's names in the movie line analogy for lists.

4.1.4 The Procedural Knowledge Model

The Procedural Knowledge Model (PKM) concerns linked lists, and was automatically constructed from initial rounds of experiments with students: a first PKM was trained on the logs from the first two versions of ChiQat-Tutor-1, ChiQat-Tutor-v1.1 and ChiQat-Tutor-v1.2, and was used by ChiQat-Tutor-v1.3 through ChiQat-Tutor-v1.5; the final PKM was trained on all the logs we obtained from the five ChiQat-Tutor-v1 versions, and is the model used by ChiQat-Tutor-v2. Details on how the PKM was trained are provided in Section 4.5.

The core of the PKM is a probabilistic graph equivalent to a Markov Chain, whose main components are *states* and *actions*.[2] A state is a snapshot of ChiQat-Tutor's virtual machine, which includes the simulated linked lists. Linked lists are internally represented with graphs. This representation allows the flexibility of modeling unusual or inconsistent linked list configurations, which can happen as a result of student exploratory actions. Actions in ChiQat-Tutor are first-class objects, which are created by the students from C++ or Java-like commands. Actions can modify a state into a different one.

The model is represented with a *simple directed graph* with two types of nodes, *state nodes* and *action nodes*, and the constraint that a state node can only point to action nodes, and that an action node must point to exactly one state node. The set of actions in the graph is associated with a probability mass function. Namely, each action is associated with the probability that a student will take that action.

Figure 4.4 shows an example of a graph for problem 1, which requires the student to insert a new node that contains "3" in the presented linked list. This graph is generated from only one student session for reading clarity, as the graphs generated with the entire dataset are much larger (see Table 4.5). "Undo", "redo" and "restart" operations are not represented in this graph. In the figure, large boxes represent states, and hexagons represent actions. Each state node contains a smaller graph representing linked lists and the variables to which some of the nodes of the list are linked. The figure also reports the "goodness" value (g) of each state, which represents a lower bound on the probability that a student traversing that state will eventually reach a correct solution; the state nodes are color coded according to goodness, red if $g = 0$, yellow if $0 < g < 1$, green if $g = 1$. The edges in the figure are annotated with an estimate of the probability that a student would traverse them while solving a problem. Another important quantity is the node *criticality* c, the probability that the student will make a fatal mistake at the following step (note that leaves in the PKM do not have criticality values, since no further action was taken in the logs along that specific path). Finally, each state includes μ_T and σ_T, namely, the mean and standard deviation of students' *think time*, i.e., the time students took to exit from that state (omitted from Figure 4.4); this timing information is used to compute a student's uncertainty, and the trigger time T to provide proactive feedback (please see Section 4.4.2).

For the interested reader, training of the PKM, including how g and c are computed, is explained in more detail in Section 4.5.

---

[2]The idea was initially inspired by the work of Barnes and Stamper, who automatically extracted Markov Decision Processes from past student interactions with their logic proof tutor [32, 33].

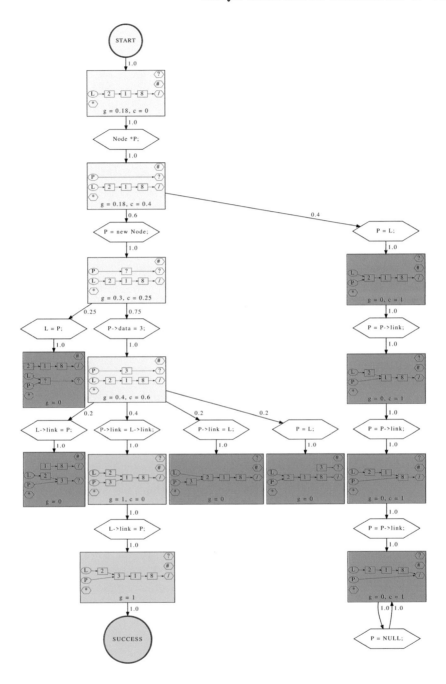

Figure 4.4 Example of a PKM graph. Large rectangles are state nodes; large hexagons are action nodes. Colors indicate the goodness (g) of a state node: red if $g = 0$, yellow if $0 < g < 1$, green if $g = 1$.

4.2 USER INTERFACE

The graphical user interface is responsible for the main interaction with the student. The interface is divided into four main views, as can be seen in the block "user interface" in Figure 4.1; in turn, they correspond to four windows in the interface, as can be seen in Figures 4.2 and 4.3, and later in this chapter, Figures 4.6, 4.7 and 4.8. The top left window (corresponding to *Problem view* in Figure 4.1) contains the description of the problem to be solved; the top middle window (*Feedback View*) is devoted to feedback messages given to the student; the bottom left window (*State Space View*) represents the current state of the data structure and finally, the fourth right window (*Student Code Input* in Figure 4.1) is the area where students can enter commands and see a history of the previously executed operations. Specifically, students enter a command at a time in the bottom "command line window", which is greyed out in Figure 4.2, and see the list of all the commands that have been successfully executed so far on the top right.

In block problems (Figure 4.3), the *State Space View* is split into as many different scenarios as are mentioned in the problem statement (two in Problem 6 and three in Problem 7). As shown in Figure 4.3, for block problems, the command line window is expanded so that the student can enter a block of code; as for step by step problems, the top of the right window shows the block that has been successfully executed.

Using this interface, students can interactively manipulate the data structures using C++ or Java commands. The command interpreter is quite resilient and tries to understand the user input even if it is slightly inaccurate. This allows the student to focus more on the semantics of statements rather than on language dependent details. We will provide additional information on the command interpreter in Section 4.4, when discussing *Syntax feedback*.

Depending on the problem type, either the effect of individual commands is reflected immediately on the graphical representation (the *State Space View*) or a block of commands is executed at once in the simulated environment, and if correct, its effects are reflected in the state of the data structures shown in the scenario windows.

Note that students cannot make any structural change to the data graphically. All meaningful changes must be done by code execution, and students cannot change any connections or content. However, students can manipulate the actual layout, i.e., drag around the nodes to rearrange them if the graphical representation becomes too "crowded" or if a new node partly occludes another, since we decided to keep the graphical visualizer relatively simple for reasons of efficiency.

4.3 A BIRD'S EYE VIEW OF ChiQat-Tutor IN ACTION

Before turning to an in-depth description of the tutor module, where we will mainly focus on ChiQat-Tutor's feedback strategies, we briefly describe how a general session with ChiQat-Tutor unfolds. This will help situate the feedback strategies that ChiQat-Tutor uses.

When a student starts working with ChiQat-Tutor, s/he is presented with a problem to solve in the top leftmost window. As we noted in Section 4.1.1, a problem is composed of a textual description and one or more initial scenarios, that are presented graphically in the bottom left window (*State Space View*). The initial scenario is integrated into the working *state space*, which includes relevant domain elements like variables and nodes. The student is asked to progressively modify the state space by interactively providing a *sequence of operations*, until the desired configuration of the data structure has been reached.

As we also noted above, to solve the problem students enter commands in the right window; if the command or set of commands (for block problems) is syntactically accurate "enough" (to be discussed below), it is simulated in the state space and its effects are mirrored in the graphical representation.

During problem solving, the student can receive a variety of feedback, to be discussed below, and in ChiQat-Tutor-v2, they can ask for an example (see *Example* button in left top window), which is played by the WOE engine; one WOE exists per problem, and its definition is provided to the engine for processing (WOE definitions are part of the domain model as described above). The WOE is shown to the students by executing actions at each step while making appropriate transitions to other steps in the WOE graph. Such actions include either displaying the next tutoring step if it consists of a verbal message, or actually executing commands for the problem and changing the state for the data structure. While each step in a WOE has a verbal component, many steps refer to a command that the human tutor would write on paper, and that ChiQat-Tutor executes, and shows at the top of the right window. For example, looking at the WOE in Table 5.6, the first five steps do not have commands associated with them. But step 6 does, since it implies that a command `Z=X-->link` should be executed.

Finally, when students think they have solved the problem, they can submit the solution (*Submit* button, in the top left window). At that point, the Solution Evaluator evaluates the solution, as we describe next. Note that students can also restart the same problem (see *restart* button in the same window), at any point while solving the problem itself.

4.3.1 Solution Evaluator

The solution evaluator is a constraint evaluator that checks the correctness of the solution. As we noted in Section 2.4, in the CBM framework, a constraint is composed of a *relevance condition* and a *satisfaction condition*. Evaluation of a constraint results in three outcomes:

1. A constraint is *irrelevant* when the relevance condition is not verified;

2. It is *satisfied* when both relevance and satisfaction conditions are verified;

3. It is *violated* when only the relevance condition, but not the satisfaction condition, is verified.

The computational evaluation of constraints is fairly simple and efficient. As a computational unit, each constraint comprises three fundamental functions: a first boolean function checks the *relevance* of the constraint with respect to the solution; a second boolean function checks the *satisfaction* of the constraint and finally, a *feedback* function returns information used to generate feedback for the student. The constraint evaluator has access to two sources of information: the current student solution and an exemplary correct solution provided by the designer. This exemplary solution, while not necessarily the only possible correct solution of the problem, allows ChiQat-Tutor to evaluate the problem-dependent properties of a student's solution, like the expected values of the final lists. As we described above in Section 4.1.2, constraints also encode properties of a correct solution / well formed list, such as that the list must be free of cycles, etc.

Overall, evaluating student solutions via a constraint-based paradigm recognizes and accepts different correct student solutions without having to list them all. This is important in a domain like data structures, where alternative procedures can be used to achieve the same results.

Once the solution evaluator has evaluated the solution, it will generate the appropriate feedback to the student (which we call *final feedback*, for reasons that will become apparent shortly). If the solution is correct, the student is congratulated on solving the problem correctly. Figure 4.3 shows one such case.

If the solution is not correct, an appropriate message is generated. First, a standard prefix is used that does not change *Uhm, there is a problem with this solution*. Then, a diagnostic message is generated, based on the information derived from the violated constraint(s). For step-by-step problems, we designed 12 different diagnostic messages, such as *One or more nodes should have been deleted*; and, *The list L has incorrect values. The list L ends with garbage*. For block problems, one example of diagnostic message (out of eight) is *In scenario #1, The list L has incorrect values. In scenario #2, The list L has incorrect values. In scenario #2, Variable P points to something wrong* (we did not focus on making these messages more concise).

ChiQat-Tutor-v1 and ChiQat-Tutor-v2 differ on whether a student is required to correctly solve a problem before moving to the next one. Specifically, in ChiQat-Tutor-v1 students could ask for a different problem by pressing the "Problems" button at any time, although few did. In ChiQat-Tutor-v2, instead, the button was disabled until they submitted a correct solution for the current problem; when that happened, they could then revisit any problems they had already solved, or move forward to the next problem.

As one would expect, even in ChiQat-Tutor-v1, students are likely to move to a new problem if they solved the previous problem correctly, and to restart it if they did not, and the majority do so; but occasionally, some move on to a new problem even if they did not submit a correct solution. Even more rarely, they repeat a problem even if they solved it correctly, which is allowed in both versions of ChiQat-Tutor (we will explore these behaviors in Section 5.3.4).

4.4 TUTOR MODULE

In ChiQat-Tutor, we mainly focused on three pedagogical strategies: how to provide and modulate feedback, and which type of feedback, to the student; and how and when to provide them with Worked Out Examples or analogies (as we will see, analogies were folded within the WOE version). The rationale behind these strategies was first presented when we discussed Modes of Learning in Section 2.1.2, and then justified by our analysis of human–human tutorial interactions in the preceding Chapter 3. This section will be devoted to the various types of feedback ChiQat-Tutor provides. Whereas conceptually and in the implementation of the Tutor Module, an example manager module does exist, its functions have just been described in Section 4.3, and hence, we will not touch on it in this section.

In ChiQat-Tutor, the type of feedback provided at each given point during problem solving is a point in a three-dimensional space:

1. Code feedback: provided when the student enters a command in the command window. It can address either syntax or executability

2. Reactive or proactive: providing feedback on moves the student has done so far or anticipating the appropriateness of future steps

3. Positive or negative: whether the provided feedback commends the student on the correctness of their step, or points out mistakes

Note that the *Final Feedback* we described above is not included in this three dimensional space, even if it obviously can be positive or negative. *Final Feedback* occurs after the student has submitted the solution, while the feedback we describe here occurs, **while** the student is solving the problem.

Given the different types of feedback, five different versions of ChiQat-Tutor-v1 were experimented with, as detailed in Table 4.1 (feedback wise, ChiQat-Tutor-v1.5 and ChiQat-Tutor-v2 provide the same types, with ChiQat-Tutor-v2 adding WOEs and analogies).

4.4.1 Code Feedback: Syntax and Executability

This kind of feedback is provided when the student enters some code, most often a single command, in the command line window (it is greyed out in all figures showing ChiQat-Tutor's interface in this chapter, other than in

Table 4.1 Feedback Types in Different Versions of ChiQat-Tutor v1

Feedback type	CQT-v1.1	CQT-v1.2	CQT-v1.3	CQT-v1.4	CQT-v1.5
Syntax	Rudimentary	Yes	Yes	Yes	Yes
Execution	Rudimentary	Yes	Yes	Yes	Yes
Reactive procedural	No	No	Yes	Yes	Yes
Proactive procedural	No	No	No	Yes but sporadic	Yes
Final	Yes	Yes	Yes	Yes	Yes

Figure 4.8). First, the command needs to be syntactically correct; if not, *Syntax feedback* is provided. If the command is syntactically correct but cannot be executed in the current state space, *Execution feedback* is provided. Code feedback, whether syntax or execution, is always negative.

As concerns syntax, the parser of ChiQat-Tutor tolerates minor syntactical imperfections, and will interpret the command in those cases (see below for an example). When the mistake is more substantial, ChiQat-Tutor matches the student's input with a set of error rules, on the basis of which appropriate messages are generated. Generating good syntax error messages, whether in ChiQat-Tutor or in actual interpreters/compilers, requires good guesses as to what the programmer actually *intends*.

ChiQat-Tutor's interpreter tokenizes text and parses token streams with respect to a grammar, just as standard parsers do. However, it produces many tokenizations and many parses, each weighted by some measure of likelihood. Valid input gets tokenized and parsed with weight zero. For invalid input, higher weighted tokenizations and parses are deemed to be less likely.

Tokenizations are generated by adding error-keywords to the set of actual keywords, and by using the standard edit distance (Damerau–Levenshtein distance) metric to find plausible interpretations accounting for typos and misspellings (please see [86], p.364, for a definition of edit distance). Though there are many potential tokenizations, the system only generates them one-by-one in order of increasing weight, until a solution is found or a threshold is reached. Tokenization steps of positive weight have error messages associated with them.

After tokenization, tokenized input is parsed according to a grammar, but the grammar includes "error rules", e.g., $pexp \rightarrow num$, which allows a number to be interpreted as a pointer expression. Each error rule has a positive weight associated with it and, just as with tokenizations, parses are generated one-at-a-time in order of increasing weight. Each error rule also has a message associated with it.

At the end of parsing, the module returns the lowest weighted tokenization and parse, provided one exists below a prescribed threshold (experimentally determined), along with an error message if that weight is non-zero. Table 4.3

Table 4.2 Examples of Different Feedback Types

Feedback type	Example
Syntax	OK, there are some problems with this input. You're trying to write a delete statement, right? "->link"" only makes sense if what comes before is a Node pointer.
Execution	You want to execute: L2->data = 5; I'm sorry, I can't do that. I can't find the node pointed to by L2.
Reactive procedural (negative)	Mmmhh... Probably you can't go very far from here... Pointer L was pointing to node 12, now it points to node 8. Node 12 was being targeted by pointer L but now it is abandoned.
Reactive procedural (positive)	Good move! Variable T was pointing to node 2, now it points to node 9.
Proactive procedural	(Shown in Figures 4.2, 4.6-4.7; described in Section 4.4.2)
Final (negative)	Uhm, there is a problem with this solution. One or more nodes should have been deleted.
Final (positive)	Good job! You have solved problem 4.

Table 4.3 Examples of Diagnostic Syntax Feedback

Prefix	Diagnosis	Specific error
OK, there are some problems with this input.	You're trying to write a delete statement, right?	Did you mean "delete" instead of "detele"?
OK, there are some problems with this input.	You're trying to write a delete statement, right?	data and link are the only things that can appear to the right of ->, they are the only Node fields.

includes examples of such messages, and shows how ChiQat-Tutor's interpreter can recognize minor syntax mistakes (the misspelling *deiele*), but also more fundamental mistakes, such as using an undeclared field.

In Table 4.4, one additional example of this type of diagnostic syntax error message is included, to compare it with what a standard compiler would show. In this case, in the input p-.link = NULL;, the student has used −. instead of the correct →. Via an alternate tokenization, "-." is interpreted as "->", with small penalty since the edit distance is small. The table shows the error messages g++, a standard C++ compiler, may return and the message ChiQat-Tutor returns (without its prefix).

If the command is syntactically incorrect, but no alternative tokenization was found below the predefined threshold, in most cases, the syntax feedback

Table 4.4 One Example of Syntax Errors and Messages

g++	error: comparison between distinct pointer types Node* and int(*) (const char*, const char*) throw () lacks a cast error: lvalue required as left operand of assignment
ChiQat-Tutor	You're trying to write a pointer assignment statement, right? Did you mean "->" instead of "-."?

returns a message with prefix *Sorry I cannot understand*, followed by the actual command, such as

```
Node *T;
T = L;
while(T != P->data){
T=L->link}
```

Other additional syntax feedback messages are returned, but they are more generic; for example, *There's a syntax error here, and I can't figure out what you were trying to do.*; these are used only when none of the others apply.

Finally, if the command is syntactically correct, it may still be the case that it cannot be executed in the current state space. In this case, *Execution feedback* is provided. An execution feedback message starts with restating what the students wants to do, then inserting the string *Sorry, I can't do that.* and then a diagnostic of why the command cannot be executed. Execution feedback can be relatively simple, as when the student is attempting to reference a variable that has not been declared:
You want to execute: L2->data = 5; *I'm sorry, I can't do that. I can't find the node pointed to by* L2.

Execution syntax can become much more complex especially in the later problems 6 and 7 when students use loops; Figure 4.5 shows an example for Problem 6.

4.4.2 Reactive & Proactive Feedback

Reactive procedural feedback is given during the problem-solving process, in reaction to student moves. *Proactive procedural feedback* is also given during the problem-solving process, but instead of reactively responding to students' past actions, ChiQat-Tutor tries to anticipate the *future* moves of a student, and possibly initiates a tutorial interaction with the student.

To generate both, we take advantage of the PKM. Reactive procedural feedback is dynamically generated by comparing the state spaces before and

You want to execute:

```
Node *M;
M = L;
while (M->link != P) {
  M = M->link;
}
T = M;
M = NULL;
delete M;
```

I'm sorry, I can't do that.
In scenario #1, I tried to execute `delete M;` *but I can't find the node pointed to by M.*
In scenario #2, I tried to execute `delete M;` *but I can't find the node pointed to by M.*

Figure 4.5 Execution Feedback for an error in Problem 6

after a student's move, and explaining the effects of that move to the student. Reactive procedural feedback can be used to correct a student's mistake (*negative* feedback) or to reinforce the understanding of correct moves (*positive* feedback).

To decide when to provide *proactive feedback*, ChiQat-Tutor monitors the student's activity. If the situation is considered *critical* (as computed based on the PKM and the state space, please see below), and enough time T has elapsed since the last student move (the trigger time T will be defined below), ChiQat-Tutor initiates the proactive interaction.

Reactive and *Proactive* feedback are clearly the more sophisticated kinds of feedback ChiQat-Tutor can provide. We did not distinguish between these two types of feedback in our tutoring dialogue analysis, meaning, we did not mark feedback along the dimension of whether the tutor is reacting to the immediate effects of the current student move, or anticipating its long-term effects, which often presupposes predicting which future actions the student may take. If you recall from the previous chapter, feedback is often a short episode, inside which both DPIs and DDIs occur. Some of the DPIs refer to potential wrong future behaviors. As we informally observed, tutors often proactively anticipate the next student move and attempt to steer them in the right direction. Sometimes, tutors use a DPI to explicitly tell students what to do; alternatively, tutors try to elicit the right move from the student using more subtle strategies, such as a hint couched as a question or prompt.

To really know whether the DPI or other tutor moves refer to the student's current step or anticipate a future step, tutors should have thought aloud (impossible in an actual tutoring session) or reflected on the tutoring session after the fact, which would have required at least a double-time commitment by the tutors (and consequently doubling the resources, monetary and otherwise, that we could devote to the tutoring dialogue collection effort). Even if that had been feasible, it is not clear that tutors would be able to make that distinction to the degree of detail required by a computational implementation.

Rather, we resorted to determining such contextual factors via the PKM, which for each state the student may find themselves in after a move, provides the *goodness* of a student's move and its *criticality*, two quantities that embody an assessment of the future consequences of the current move: recall that g represents a lower bound on the probability that a student traversing that state will eventually reach a correct solution, whereas c represents the probability that the student will make a fatal mistake at the following step. Additionally, the PKM will also allow us to estimate the level of *uncertainty* with which the student made that move. It is important to note that using the PKM to assess these quantities, corresponds to dynamically computing the current snapshot of a student model, that as we mentioned above, does not exist in a declarative form in ChiQat-Tutor.

Specifically, the conditions under which reactive and proactive feedback are provided are as follows, where g is goodness and c is criticality; s is the previous state and s' the current state. For conditions 2 and 3, the addition of uncertainty and trigger time, respectively, allows ChiQat-Tutor to modulate both positive and negative feedback, since neither do tutors comment on every wrong move nor do they commend every correct move:

1. $g(s') = 0$: reactive negative feedback;

2. $g(s') > g(s) \wedge uncertainty > threshold$: reactive positive feedback;

3. $g(s') > 0 \wedge c(s') > 0$ and trigger time T has been reached: proactive feedback

Both goodness g and criticality c of a state in the PKM are computed a priori when the PKM is trained (please see Section 4.5 for details); in real time, the current state space is compared to the states in the graph, and the values of the matching state in the graph are used. The computation of uncertainty is discussed below, and so is the computation of exactly when in the interaction to provide proactive feedback, since many states satisfy the $g > 0 \wedge c > 0$ condition. In fact, it is possible that for the same current state s' a student receives reactive positive feedback first, and then proactive feedback as well: reactive feedback is provided by comparing the goodness of the current state s' to that of the previous state s; if this same student then spends too much time in s' and the trigger time T is reached, proactive feedback will be provided as well.

4.4.2.1 Reactive Procedural Feedback

This type of feedback is provided in the following circumstances. If the student just got into a hopeless state, i.e., a state with goodness $g = 0$, then a *negative feedback* message is generated, to help the student recover from the error (recall that $g = 0$ means that the estimated probability of subsequently reaching a solution is zero). If the student has made a good move, i.e., has improved his/her probability of reaching a correct solution, *and* the student showed uncertainty, then a *positive feedback* message is provided. As we discussed earlier in Section 2.1.2, one function of positive feedback is uncertainty reduction, in that students perform correct moves tentatively or even randomly, and it is important to strengthen correct knowledge that the student may not have fully acquired yet.

To estimate uncertainty, ChiQat-Tutor keeps track of the time taken by the student to perform the step, and of whether the student had performed an *undo/redo/restart* while solving problem i since starting the session. Either can be used to estimate uncertainty. Specifically, the student is deemed to show uncertainty if the time s/he has taken is larger, by more than a standard deviation, than the average time used by past students at that same point (as recorded in the PKM as the *think time* associated with a state node), after applying a correction factor based on a student's personal history (each individual student may be faster or slower than the average). Alternatively (or additionally, since the two conditions are not exclusive of each other), the student is considered uncertain if s/he has performed an "undo", "redo" or "restart" operation while solving problem i.

Reactive feedback has two main parts. The first component expresses the literal negative or positive feedback, namely, refers to the goodness of the student's move; two fixed expressions are used *Good move!* (positive feedback) and *Mmmhh... Probably you can't go very far from here* (negative feedback). This fixed expression is followed by an explanation of the effects of that move on the problem state space, such as *The node that contained 7 now contains 0* (positive) or *Pointer L2 was pointing to node 5, now it points to node 3. Node 5 was being targeted by variable L2 but now it is abandoned* (negative). To create the explanation, ChiQat-Tutor compares the previous state with the current state, then highlights the differences between those states. The facts to be communicated are chosen using a set of rules. Finally, the actual sentence (*surface realization* in Natural Language Generation parlance [148]) is produced using the SimpleNLG library [350].

4.4.2.2 Proactive Procedural Feedback

Given our informal observations of proactive behavior on the part of our tutors, the strategy implemented in ChiQat-Tutor consists of a tutor–student interaction which is comprised of the three components. We will illustrate them by referring to Figures 4.2, 4.6 and its alternative 4.7, that represent ChiQat-Tutor's reaction to one correct and one incorrect student's answer.

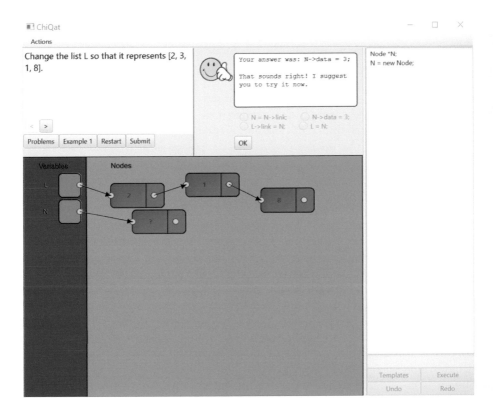

Figure 4.6 Proactive feedback, third component: reacting to correct student's choice.

Proactive procedural feedback procedes as follows

1. A question from the tutor, in three parts, as illustrated in Figure 4.2:

   (a) A statement of the goal to be accomplished by the next move;

   (b) An explicit question about how to achieve that goal;

   (c) Up to four possible moves, which include the correct answer and some of the most frequent incorrect answers given by students in our human–human tutoring dialogues.

2. A reply by the student, in the form of choosing one of the alternatives by clicking on it.

3. Feedback from the tutor, as illustrated in Figures 4.6 and 4.7. In Figure 4.6, the tutor provides positive feedback, and suggests that the student tries that command. If instead the student chooses a wrong alternative, ChiQat-Tutor provides appropriate negative feedback, as shown

in Figure 4.7. If the students were to scroll down in the widow containing the feedback message, they would find the following explanation: *Here is what will happen if you do what you suggested. Pointer L is now pointing to node 2, then it will point to a node with undefined content. Node 2 is being now targeted by pointer L but it will be abandoned.*

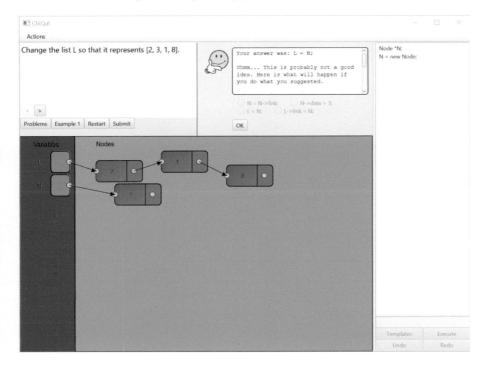

Figure 4.7 Proactive feedback, third component: reacting to wrong student's choice.

Once the proactive feedback interaction starts, the three components must occur, and specifically, the student must choose one of the proposed moves before s/he can continue working on the problem. When the tutor provides the last part of the feedback, whether positive or negative, the ChiQat-Tutor's interface reverts to enabling the student's input (note that in Figures 4.2, 4.6 and 4.7 the bottom right command line window is greyed out). Figure 4.8 shows the positive reactive feedback that is provided to this particular student after they actually input the suggested command, presumably following after Figure 4.6. Note that this happens because of the conditions that trigger positive reactive feedback and that we discussed earlier, namely, $g(current\ state) > g(previous\ state) \land uncertainty > threshold$; reactive feedback is independent from whether proactive feedback was provided just prior to the current student move. As it happens, this specific student's behavior prior and throughout the proactive feedback episode must have exhibited enough uncertainty that reactive positive feedback was triggered.

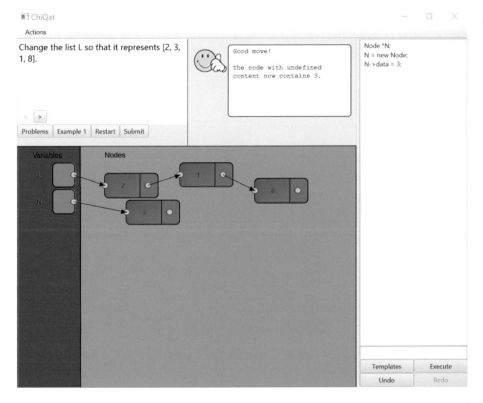

Figure 4.8 After proactive feedback: reacting to student's command.

As we noted above, a necessary condition for a proactive interaction to be initiated is based on goodness and criticality, and specifically, $g > 0$ and $c > 0$: this means that the current state is not hopeless ($g = 0$) but somewhat critical. We determined the threshold $c = 0$ for criticality by simulating the effects of different threshold values on our training data.

Additionally, we need to determine the specific trigget time T to initiate a proactive interaction, since many states will satisfy that condition. T is in seconds, it is computed for each state with $g > 0 \ and \ c > 0$, and the computation is reset (namely, proactive feedback does not occur) if the student does a move while T is being computed.

The trigger time T is computed as in the following formula:

$$T = T_{min} + (T_{max} - T_{min})B(1 - c) \qquad T_{min} = \max(5, \mu_T - \sigma_T)$$
$$T_{max} = \max(5, \mu_T + \sigma_T)$$

T is affected by four variables: mean μ_T and standard deviation σ_T of the time that previous students spent in the current state (the *think time* associated with each state node in the PKM); the criticality of the current state c and a measure of the correctness of the current student's behavior B, to be defined below. The constant 5 (in seconds) in the computations of T_{min} and T_{max} is an educated guess for a reasonable minimum think time, namely, we allow students to think for at least 5 seconds before interrupting them. Essentially, this formula computes a time T between plus and minus a standard deviation from the mean student think time in the current state, modulated by the current student's behavior B, where $0 \le B \le 1$. If the student has a generally good behavior (B closer to 1), T will be longer; conversely, T will be shorter if the student has a generally worse behavior (B closer to 0).

In turn, B is computed as a smoothed function of how often the student visits "good" or "hopeless" states (n is the current step, $n-1$ is the previous step, g is the goodness of the current state):

$$B_n = \alpha x + (1-\alpha)B_{n-1} \qquad \alpha = 0.5 \qquad x = \begin{cases} 0 & \text{if } g = 0 \\ 1 & \text{if } g > 0 \end{cases}$$

4.5 TRAINING THE PKM GRAPHS

We conclude the discussion of the conceptual architecture by discussing how the PKM graphs are trained, one PKM graph per problem (this section is included for the readers who are curious about the specifics of how we obtained the PKM graphs, but can be safely skipped otherwise).

Recall that the PKM is a probabilistic graph, composed of state nodes (represented by boxes) and action nodes (represented by hexagons), please see Figure 4.4; a state node can only point to action nodes, and an action node must point to exactly one state node. A state node is a snapshot of ChiQat-Tutor's virtual machine, which includes the simulated linked lists. Actions correspond to student C++ or Java-like commands, and can modify a state into a different one.

To build the PKM, we start from the logs in ChiQat-Tutor. First, the student actions from the log file are executed. For each action, a new state is generated. If the new state s' matches a state s already included in the PKM, the frequency of the pre-existing s is increased; if not, s' is added to the PKM. Actions are processed similarly. Matching states and actions to those already existing in the PKM decreases the size of the graph and helps reduce the amount of overfitting.

The algorithm to match states and actions is complex since the match between states must be based on the structure of the list and semantics (i.e. what the state represents, not surface differences like what name a temporary variable has), and in turn, a state is a graph in itself. The matching algorithm looks for isomorphism relations between two states. In our case, isomorphism relations are highly constrained by the specific structure of these graphs, which makes the search manageable (see below for some discussion of the complexity of our algorithm). In particular, components of states that are deemed "interchangeable" when checking for isomorphism are temporary variables and data nodes with identical values. For each state s, ChiQat-Tutor keeps track of the matching history, namely, the sequence of previous states from which s derived; if more than one isomorphism relation is found, namely, more than one existing state matches the new state s' under consideration, the sequence of states from which each existing state derived is analyzed in order to choose the state s that truly matches s'.

After a state is added to the PKM (which means that it is a new state that did not exist before), further processing occurs. First, ChiQat-Tutor checks if that state represents a correct solution for the current problem, in which case, the state is linked to a special "success" node. Second, a state which is not linked to "success" gets annotated with statistics about the time students took to exit from that state, what we call *think time*. This information is used by ChiQat-Tutor to assess student uncertainty, as we described earlier.

Individual students' histories are recorded into separate structures; additionally, these structures include sets of mappings that map the real states from the logs in ChiQat-Tutor to the matched states in the graph, as well as real actions to matched actions. Namely, ChiQat-Tutor keeps a history of all the original student information together with mappings to the global graph.

After the graph has been built, the frequencies associated with states and actions are turned into probabilities using maximum likelihood estimation. These probabilities are stored with the edges of the graph.

Then, two important measures are computed. The first is a "goodness" value (g) for each state of the graph: g represents a lower bound on the probability that a student reaching that state will manage to solve the problem. A state with $g = 0$ is deemed "hopeless", i.e., it is not possible to come to a solution from such a state. g is computed by summing the probabilities of the k most likely paths (with k empirically set to 10) from the current state to the special "success" node to which all the correct states are linked.

The second measure is the "criticality" (c) of each node. The criticality of a node is the probability that the next action from the current state will result in a hopeless state ($g = 0$). c is computed by summing the probabilities of all the outgoing actions that immediately lead to a state with $g = 0$. For example, the fourth state in the left branch in Figure 4.4 has $c = 0.6$ since three outgoing actions immediately lead to a state with $g = 0$; each of these actions has probability $= 0.2$.

At run-time, a new student's actions are mapped to the PKM. The comparison between the student's behavior and the model allows ChiQat-Tutor to provide feedback, as discussed earlier in this chapter.

Table 4.5 describes the size of the PKM, in terms of number of state nodes and action nodes, for each of the five step-by-step problems in the curriculum, whereas the column *Actions* under "Data" describe the number of actions performed by students per problem, in those sessions. The PKMs in this table

Table 4.5 Procedural Knowledge Models: size of Model by Problem

Problem	Data		PKM	
	Sessions	Actions	State nodes	Action nodes
Problem 1	259	2160	295	399
Problem 2	267	4130	1034	1363
Problem 3	236	2033	314	460
Problem 4	184	2037	549	721
Problem 5	150	1724	323	424

were obtained at the transition from ChiQat-Tutor-v1 to ChiQat-Tutor-v2, namely, on all the log data that had been collected with ChiQat-Tutor-v1: these are the PKMs used by ChiQat-Tutor-v2.

The attentive reader may ask: since starting with ChiQat-Tutor-v1.3, some versions of ChiQat-Tutor-v1 did use reactive/proactive feedback, which are triggered according to goodness and criticality of nodes in the PKM, which PKMs were available then? We trained a first set of PKMs based only on ChiQat-Tutor-v1.1 and v1.2, which were used by the remaining three versions of ChiQat-Tutor-v1. These initial PKMs were smaller than the PKMs in the final set, but not by a large amount: in the final set of PKMs, the number of state nodes increased by a minimum of 8% (problem 1) to a maximum of 35% (problem 4), and the number of actions from a minimum of 12% (problem 1) to a maximum of 35% (problem 4).

The computational complexity of building the PKM, fully discussed in [131], is briefly summarized here. We will note that in practice, the construction of the PKM is very fast. With the data set currently available, the entire model can be built in a few minutes even on a modest personal computer. For the interested reader, [131] discusses the learning curve of the PKM in full detail.

The construction of the PKM comprises two main operations: the matching operation that compares two state spaces to build the graph itself, and a shortest-path algorithm to traverse the final graph and compute g and c.

The first phase is potentially exponential, as it concerns checking for isomorphism relations between graphs.[3] However, in practice, the size of the graphs is limited, since students work on problems involving only small lists; these problems result in small and structurally constrained graph representations that can quickly be compared. If N is the number of student actions recorded in the log files, the worst case time complexity of the first phase of the graph construction is $O(N^2)$, as each new node could be potentially be checked for equivalence with every other node in the graph.

---

[3]The exact complexity of the problem of graph isomorphism is still unsolved.

After the graph has been computed, to calculate the various probabilities and measures discussed above, we need to traverse the graph, and specifically, compute shortest path, for which we use the Bellman–Ford algorithm. Its worst case complexity is $O(mn)$, where m is the number of edges and n is the number of nodes. Given our model, $m = n$, because both action and states are nodes, each action can be linked to only one state; each state can be linked to multiple action nodes, but note that action a_i can only be reached by one immediately preceding state (for example, in Figure 4.4, multiple `P=P->link;` nodes appear in the right branch, but each of them is only reachable by the previous state node, not by others). In the worst case, $n = 2N$, as each student action could result in a new action node and a new state node. However, in practice n is much smaller than $2N$, as most of the students' actions and the resulting state spaces can be matched to nodes already added to the PKM during construction.

CHAPTER 5

Evaluation in the Classroom

With Rachel Harsley

Google

With Stellan Ohlsson

University of Illinois at Chicago

CONTENTS

U SER evaluation is crucial to assessing whether an ITS is effective. Effectiveness most often concerns learning; however, the students' attitudes toward the ITS are often measured as well. The evaluation we ran directly builds on our findings from the human–human studies described in Chapter 3. As summarized in Table 3.19 and in Section 3.5, at a high level, the pedagogical strategies that impacted student learning the most were feedback, whether by itself or embedded in WOEs; features of WOEs such as number of utterances in a WOE; presence of analogy, and dialogue acts within analogy (other than for stacks). Hence, in the following, we will focus on these strategies. However, we will depart from our human–human tutoring findings in two ways: first, all evaluations we will discuss in this chapter pertain to students learning lists, since only problems on this data structure were fully developed in ChiQat-Tutor; likewise, we will assess these three strategies in a holistic way, since ChiQat-Tutor is not able to engage in a conversation with its users yet. Namely, we cannot evaluate models in which we consider sequences of dialogue acts, as in the significant models of type 2 discussed in Section 3.4.2.

5.1 EVALUATION METRICS

Our main evaluation metric will be in terms of learning gains; we will also discuss measures such as time on task, number of problems completed and several other features extracted from the logs of ChiQat-Tutor. For ChiQat-Tutor-v1, we will also discuss the result of a user satisfaction survey.

A reminder that we use absolute learning gain (the difference between the pre- and post-test scores) as our measure of learning (please see discussion in Section 3.1.1). As we noted earlier, absolute learning gain is appropriate as a measure, when there are no differences among pre-test scores among experimental groups or conditions. As for the human tutored conditions, also for the evaluation of ChiQat-Tutor, we ran several tests on whether the pre-tests among experimental groups and/or conditions significantly differ (please see Sections 5.2, 5.3.2, and 5.3.3). Either we found they did not or when they did, as detailed in Section 5.3.2, the difference was due to the background of the students in those classes (being enrolled in a required class for majors, or for non majors). In this latter case, the solution was not, say, to adopt normalized gain as opposed to absolute gain, but rather, to separately analyze the two groups of students, as we did.

5.2 LEARNING WITH PROACTIVE AND REACTIVE FEEDBACK

We evaluated ChiQat-Tutor-v1 at both UIC and at the United States Naval Academy (USNA). Both the types of students were taking an introductory data structure class, and interacted with the system in a laboratory section shortly after having been presented linked lists (presumably for the first time, given the introductory nature of these classes). The lab session lasted 1 hour 15 minutes to 1 hour 30 minutes long. We asked them to complete a pre-test, work with ChiQat-Tutor-v1, complete an identical post-test, and finally fill in a survey. In ChiQat-Tutor-v1.1, both tests and the survey were on paper; additionally, the instructor was in charge of giving the student a short

Table 5.1 Learning Gains of Students

Condition	N	Pre-test		Post-test		Gain	
		μ	σ	μ	σ	μ	σ
Control	53 (UIC)	.34	.22	.35	.23	.01	.15
ChiQat-v1.1	61 (28 UIC, 33 USNA)	.41	.23	.49	.27	.08	.14
ChiQat-v1.2	56 (UIC)	.31	.17	.41	.23	.10	.17
ChiQat-v1.3	19 (USNA)	.52	.29	.65	.26	.13	.24
ChiQat-v1.4	53 (UIC)	.53	.24	.63	.22	.10	.16
ChiQat-v1.5	30 (USNA)	.37	.24	.51	.26	.14	.17
Human Tutoring	54 (UIC)	.40	.26	.54	.26	.14	.25

tutorial on the system by solving the first problem of the curriculum in front of them. From ChiQat-Tutor-v1.2 on, tests and survey were online, and so was a brief written tutorial providing some minimal information on the system and demonstrating how to solve the same first problem. Notwithstanding the medium of presentation (hardcopy or digital), the pre/post-test were identical and did not change across conditions; in fact, the pre-/post-test was the same as the test that students took in the human condition. Pre-/post-tests can be found in Appendix B.

To measure student learning, Table 5.1 shows the results of our between-subject study, in terms of learning gains of seven groups of students. In five conditions, each student interacted with one version of ChiQat-Tutor-v1. The control condition is the same that we discussed in Section 3.1 and we compared to the human tutored conditions (as a reminder, the 53 students in the control condition did not receive any form of instruction between the two tests: they attended a lecture about an unrelated topic, on discrete mathematics). As our ceiling, we also compare learning gains of students who interacted with ChiQat-Tutor-v1 with those of the students who interacted with the human tutors, as described in Chapter 3.

First, paired samples t-tests revealed that post-test scores are *significantly higher* than pre-test scores for each ChiQat-Tutor-v1 version (recall from Ch. 3 that students in the human tutored group, but not in the control group, also achieved significant learning on lists). ANOVA revealed an overall significant difference among the seven groups ($F(6, 319) = 3.04$, $p = .007$). Tukey post-hoc tests revealed significant differences only between the control group and the human tutored group ($p = .004$), and between the control group and ChiQat-Tutor-v1.5 ($p = 0.021$). The progression of effect sizes through the multiple versions of ChiQat-Tutor-v1 indicates a positive performance trend, even if pairwise differences are not significant. The reader may question whether the results can be trusted since the various versions were evaluated by students at different institutions. If anything, we believe that this strengthens our results since these students were studying the same material with different teachers, within a different curriculum with potential differences in course content, prerequisites, etc. In the only case in which students from both institutions interacted with the same version of the system (ChiQat-Tutor-v1.1), there was no significant difference between the pre-tests of the two groups ($t = -0.86419, df = 52.173, p = 0.3914$).

It is very promising to us that the learning gains obtained by students in the human tutoring condition and in ChiQat-Tutor-v1.5, the most sophisticated version of ChiQat-Tutor-v1, are identical.

5.2.1 Insights on Learning from Student Behavior and Perceptions of ChiQat-Tutor-v1

We performed several further analyses to understand how learning is affected by both the features of the student–system interaction, and by the student's satisfaction and perception of the system's interaction. The latter was assessed via surveys, whose analysis is discussed in section 5.2.1.2.

5.2.1.1 Student Behavior

The interaction between ChiQat-Tutor and the students is comprehensively logged. From the log files, we extracted several features and compared them using ANOVA and linear regression. We highlight the impact of *problem solved*, *path goodness* and *positive versus negative feedback*.

Number of problems solved. The problems included in the ChiQat-Tutor's curriculum are of increasing difficulty, as can be seen from the success rate for each problem (Figure 5.1). ANOVA revealed overall significant differences among the five groups on the number of problems successfully solved by the students ($F(4, 214) = 19.5$, $p < .001$). Tukey post-hoc tests showed that all pairs are significantly different ($p < .05$), other than ChiQat-Tutor.v1.1/ChiQat-Tutor.v1.4 and ChiQat-Tutor.v1.3/ChiQat-Tutor.v1.5. We will come back to this lack of difference below, when we discuss path goodness. The three groups that worked with a version of ChiQat-Tutor enhanced with procedural feedback (ChiQat-Tutor.v1.3, ChiQat-Tutor.v1.4, and ChiQat-Tutor.v1.5) generally solved more problems. Linear regression showed a small, but significant positive correlation between the number of problems solved and learning: generally, students that solved more problems also learned more ($N = 219$, $\beta = .31$, $t = 4.73$, $R^2 = .09$, $p < .001$).

Path goodness. Using the PKM, we can assess the goodness of entire student paths, defined as the average goodness of the states visited by a student while solving a problem. If procedural feedback has an effect, we would see an increase in the average path goodness from the first two versions of ChiQat-Tutor-v1, which did not provide procedural feedback, and the last three, which provided procedural feedback of varying degree. Indeed,

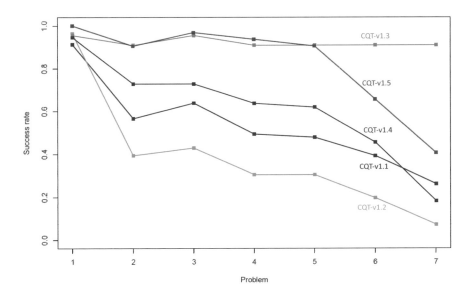

Figure 5.1 Success rates per problem, per system.

we observed such an increase (for ChiQat-Tutor.v1.1 and ChiQat-Tutor.v1.2 combined, $\mu = .19, \sigma = .13$; ChiQat-Tutor.v1.3+4+5, $\mu = .31, \sigma = .11$). ANOVA found overall significant differences ($F(4, 214) = 19.1, p < .001$). Tukey post-hoc tests confirmed all the pairwise differences ($p < .05$), except ChiQat-Tutor.v1.3/ChiQat-Tutor.v1.5 and ChiQat-Tutor.v1.4/ChiQat-Tutor.v1.5 which are not significantly different; ChiQat-Tutor.v1.1/ChiQat-Tutor.v1.4 which are marginally different ($p = .089$). As we already noted above for number of problems solved, we find a surprising lack of difference between ChiQat-Tutor-v1.3 (that does not have proactive feedback) and ChiQat-Tutor-v1.5 (which does). This difference may actually be due to the fact that students assigned to ChiQat-Tutor-v1.3 experienced a severe crash of the laboratory network that forced the students to repeat the experiment with ChiQat-Tutor-v1.3 about a week later. Whereas the crash occurred early in the session, most students had already started solving problems when the network crashed; namely, when they came back to ChiQat-Tutor-v1.3 in the second session, students had already interacted with the system (their better performance did not better dispose them toward the system, as we will discuss below in Section 5.2.1.2). The second somewhat surprising result is the strong showing of ChiQat-Tutor-v1.1, the simplest version of all: one would expect that students perform worse with it than with any other version, on all dimensions. However, it is indistinguishable from ChiQat-Tutor-v1.2 as concerns path goodness; more surprisingly, it is indistinguishable from ChiQat-Tutor-v1.4 as concerns number of problem solved, and only marginally different as concerns path goodness. We do not have an intuition as to why this may be

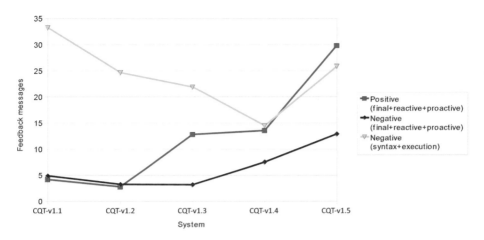

Figure 5.2 Number of feedback messages, grouped by type, in the 5 versions of the system.

the case. Linear regression found a small, but significant positive correlation between path goodness and learning ($N = 219$, $\beta = .28$, $t = 4.30$, $R^2 = .08$, $p < .001$). This finding supports the validity of the PKM and the strategy we adopted in ChiQat-Tutor, which guides the students toward productive paths.

Positive and negative feedback. The PKM allows ChiQat-Tutor to provide more feedback, in particular positive feedback, in addition to the relatively more primitive syntax, execution and final feedback (Table 4.1). This makes the latest three versions of ChiQat-Tutor-v1 able to cover more learning opportunities than the first two versions of the system. Figure 5.2 illustrates the average number of different types of messages that are provided by the five versions of ChiQat-Tutor-v1. Recall that ChiQat-Tutor.v1.1 and ChiQat-Tutor.v1.2 cannot provide reactive or proactive feedback of any sort, they only provide syntax and execution (both are always negative, because they are in reaction to a mistake), and *final* feedback, which can be positive (the student has solved the problem), or negative (the student has not solved the problem). This is why in Figure 5.2, negative feedback is grouped into two types: *syntax + execution*, and *final + reactive + proactive*.

The different groups of students received significantly different amounts of feedback of every type. We ran ANOVAs among the five conditions for the three types of feedback illustrated in Figure 5.2: each ANOVA has overall $P < .001$, pairwise differences as obtained via post-hoc tests vary.

As concerns negative feedback on syntax and execution, it progressively decreases from ChiQat-Tutor.v1.1 to ChiQat-Tutor.v1.4, but it increases back in ChiQat-Tutor.v1.5. We do not have a convincing explanation for this finding: apparently the last group of students was more prone to syntax errors. As concerns the second type of negative feedback, we can see that it progressively increases from ChiQat-Tutor.v1.1 to ChiQat-Tutor.v1.5. The increase is due to increasingly generating more reactive and proactive negative feedback messages – recall that ChiQat-Tutor.v1.3 does not generate proactive feedback, but it still generates fewer reactive negative feedback messages than ChiQat-Tutor.v1.4 and ChiQat-Tutor.v1.5 (please refer to [128], specifically Table XIII, for further details).

As concerns positive feedback, the number of positive final feedback messages is basically constant across the five versions of ChiQat-Tutor-v1, since it is almost equal to the number of problems solved by the students: usually a student receives only a single "good job, you have solved the problem" message before moving to the next problem. The amount of positive procedural (reactive + proactive) feedback greatly increases in the most recent versions of the system. In ChiQat-Tutor.v1.5, the ratio of positive procedural feedback over negative procedural feedback is approximately 2.5 to 1.

Our expectation was that the higher number of reactive/proactive messages generated in ChiQat-Tutor-v1.4 and ChiQat-Tutor-v1.5, would positively affect learning. However, linear regression did not find any significant correlation between the number of feedback messages of various types[1] and learning gain. We believe that an analysis focused on different categories of students (e.g., exhibiting more or less uncertainty) may uncover effects of the different types of feedback. Whereas we did not pursue such an analysis in this specific case, we will as concerns WOEs and students with different initial knowledge.

5.2.1.2 Student Satisfaction

After interacting with all versions of ChiQat-Tutor-v1, students were administered a survey to assess their satisfaction with the system. The survey included seven 5-point Likert scaled questions, plus an open-ended question asking them for general comments. Mean and standard deviation of the numeric answers are reported in Table 5.2.

Overall, students liked the system: they found working with any version (Q2) equally interesting, and found the feedback equally difficult (Q4). For all the other 5 questions, ANOVA found significant differences among systems (please find ANOVA values in the footnote.[2])

For these 5 questions, we ran post-hoc Tukey tests. Table 5.3 summarizes the results. A non-empty cell should be read as follows: CQ_i (row) is better than CQ_j (column) according to $Question_k$ with significance $p = 0.0xxx$

---

[1]We excluded the positive final messages, because they are basically equivalent to the number of problems solved; the total number is slightly higher, because very occasionally a student goes back to solving an already solved problem again.

[2]Q1:$F(4, 209) = 4.44$, $P = .002$; Q3: $F(4, 209) = 3.56$, $P = .008$; Q5: $F(4, 209) = 4.46$, $P = .002$; Q6: $F(4, 209) = 4.27$, $P = .002$; Q7: $F(4, 209) = 5.33$, $P = .0004$.

Table 5.2 Survey, Scaled Questions (1=No to 5=Yes)

Question	CQT-v1.1		CQT-v1.2		CQT-v1.3		CQT-v1.4		CQT-v1.5	
	μ	σ	μ	σ	μ	σ	μ	σ	μ	σ
1. Did ChiQat-Tutor help you learn about linked lists?	2.9	1.1	2.9	1.1	3.2	1.4	3.3	1.4	3.9	1.2
2. Do you feel that working with ChiQat-Tutor was interesting?	4.0	1.1	3.8	1.1	3.3	1.3	3.8	1.2	4.0	1.2
3. Did you read the verbal feedback the system provided?	4.1	1.1	4.1	1.0	3.5	1.5	4.3	1.1	4.6	0.8
4. Did you have any difficulty understanding the feedback?	2.8	1.5	2.9	1.2	3.5	1.2	2.9	1.4	3.4	1.3
5. Did you find the feedback useful?	2.6	1.2	3.0	1.0	2.8	1.4	3.4	1.1	3.4	1.2
6. Did you ever find the feedback misleading?	2.3	1.3	2.4	1.1	3.2	1.4	2.8	1.4	3.2	1.3
7. Did you find the feedback repetitive?	3.8	1.2	3.1	1.1	4.2	1.0	3.2	1.4	3.2	1.3

(* indicates statistical significance, and its absence, marginal significance). The greyed out cells on the diagonal are irrelevant, since they pit one version against itself. As expected, we found that ChiQat-Tutor-v1.4 and ChiQat-Tutor-v1.5 mostly outscored the earlier systems according to Question 1 (effectiveness in teaching linked lists), Question 3 (frequency of reading feedback), Question 5 (usefulness of feedback) and Question 7 (repetitiveness of feedback).

A surprising result concerns Question 6, whether *feedback was ever misleading*: ChiQat-Tutor-v1.5's feedback is considered more misleading than both ChiQat-Tutor-v1.1 and ChiQat-Tutor-v1.2. It is not clear why the feedback may be considered misleading, since ChiQat-Tutor does not communicate incorrect information. The feedback provided by ChiQat-Tutor-v1.1 and ChiQat-Tutor-v1.2 is minimal, so perhaps in that it is very clear; on the other hand, students do not find the feedback provided by ChiQat-Tutor-v1.4 more misleading. A confusing factor may be that the meaning of the scale is reversed (lower values are better) for Question 6 (but that would apply also to Questions 4 and 7). Retrospectively, we should have rephrased the survey to make all scales consistent.

Across all questions, ChiQat-Tutor-v1.3 fares the worst. On no question is it considered better than any other system (empty row); on the other hand, every other version is better than ChiQat-Tutor-v1.3 for at least one question (each cell in column ChiQat-Tutor-v1.3 has at least one entry). This result is most likely due to the severe crash of the laboratory network that forced the students to repeat the experiment with ChiQat-Tutor-v1.3 about a week later, and that we had mentioned earlier. This incident highlights that the acceptance of a tutoring system can be affected by external factors, which is a concern when migrating from a controlled experimental setting to the real world.

Table 5.3 Relationships between Survey Aspects

	CQT-v1.1	CQT-v1.2	CQT-v1.3	CQT-v1.4	CQT-v1.5
CQT-v1.1			Q6 $(p = 0.097)$		Q6 $(*p = 0.01)$
CQT-v1.2	Q7 $(p = 0.01)$		Q7 $(*p = 0.008)$		Q6 $(*p = 0.02)$
CQT-v1.3					
CQT-v1.4	Q5 $(*p = 0.004)$ Q7 $(*p = 0.04)$		Q3 $(*p = 0.04)$ Q7 $(*p = 0.02)$		
CQT-v1.5	Q1 $(*p = 0.003)$ Q5 $(*p = 0.01)$	Q1 $(*p = 0.001)$	Q3 $(*p = 0.006)$ Q7 $(p = 0.06)$		

We also ran linear regressions to uncover the correlation between learning gain and user satisfaction (broadly construed via these seven questions). We found a positive correlation between Question 1 (system helpfulness) and learning ($R^2 = .08$, $\beta = .29$, $F(1, 210) = 18.7$, $P < .001$); a positive correlation between Question 5 (feedback usefulness) and learning ($R^2 = .02$, $\beta = .15$, $F(1, 210) = 4.94$, $P = .027$) and a negative correlation between Question 7 (feedback repetitiveness) and learning ($R^2 = .05$, $\beta = -.23$, $F(1, 210) = 11.3$, $P < .001$).

5.2.2 Chiqat-Tutor, Version 1: Summary of Findings

Our evaluation of ChiQat-Tutor-v1 focused on the effect of various feedback strategies on learning, both via direct measures (learning gain) and indirect measures (number of problems solved, path goodness), and on student satisfaction. The conclusion is that students learn when using any version of ChiQat-Tutor, and that we observe a progression in learning gains through the various versions, which culminates in ChiQat-Tutor-v1.5 being significantly different from control. Additionally, the triad ChiQat-Tutor-v1.3, v1.4, and v1.5 in general engender better performance from the students as concerns number of problems solved, and goodness of their paths to solutions. These three versions of ChiQat-Tutor provide procedural feedback (reactive and proactive) to students; however, we did not find a direct relation between different types of feedback and students' learning, since regressions of feedback types with learning gains as dependent variable were not significant. As concerns students' satisfaction, they were mostly appreciative of ChiQat-Tutor; the system that fared the worst from this point of view was ChiQat-Tutor-v1.3, for the reasons we discussed earlier (network crash). We now turn to ChiQat-Tutor-v2.

5.3 LEARNING WITH WORKED-OUT EXAMPLES AND ANALOGY

ChiQat-Tutor-v2 builds on ChiQat-Tutor-v1.5. All feedback mechanisms were maintained; additionally, worked-out examples (WOE) and analogy were deployed, and evaluated in a controlled fashion, resulting in seven different conditions being run over the span of four semesters (two academic years). Table 5.4 summarizes all the conditions and how many subjects in each conditions we ran; Table 5.5 provides a high bird view of learning gains in all conditions (as usual, we use standard gain).

The *NoWOE* condition is basically a repetition of the ChiQat.v1.5 evaluation, since there was no change in the functionality of the system; however, the noWOE condition was run about five years later than the original ChiQat-Tutor-v1.5 evaluation. Hence, even if the population of students was similar, we cannot exclude that any differences may be due to updated curricula students were following in their course of study, students' broader exposure to Computer Science via mobile technology and social media, and the like.

We will briefly discuss the conditions in Tables 5.4 and 5.5. We will provide further rationale and discussion when we analyze specific results for specific conditions. All conditions other than the noWOE condition include one example on-demand for each problem. Hence, the student can ask for an example if they wish, but they do not have to – the example is provided if they click the Example button in the top left window in the ChiQat-Tutor interface, as shown in all the figures of the system in Chapter 4 (which are in fact images of ChiQat-Tutor-v2, since WOEs were not available in ChiQat-Tutor-v1).

5.3.1 WOE and Analogy Conditions

We now look at different types of worked-out examples that vary in their length and usage of analogy.

TABLE 5.4   Count of Samples over All Conditions

Semester (Course)	No WOE	Short WOE	Standard	Long WOE No Exit	Time Out	Initial Analogy	as Analogy	Total
F14 (211)	28		25					53
F14 (201)	16		11					27
S15 (211)	8		27			21	19	75
S15 (201)		5	3					8
F15 (211)	1	23	31					55
F15 (201)		17	18					35
S16 (211)				28	32			60
Total	56	45	115	28	32	21	19	313

TABLE 5.5   Average Learning Gain per Group and Condition

Semester (Course)	No WOE	Short WOE	Standard	No Exit	Time Out	Initial Analogy	as Analogy
				Long WOE			
F14 (211)	.08		.00				
F14 (201)	.14		.24				
S15 (211)	.16		.11			.11	.11
S15 (201)		.15	0				
F15 (211)	0	.13	.08				
F15 (201)		.11	.15				
S16 (211)				.02	.05		
All	.11	.12	.09	.02	.05	.11	.11

5.3.1.1   Standard WOEs

Long WOEs, in their *Standard* form, are the condition we experimented with the most, as the reader can observe from Table 5.4. Standard WOEs are directly informed by the WOEs we had studied in the human tutoring dialogues and include the patterns we had observed in those; for example, in WOEs, human tutors include wrong steps, that they correct right away (for instance, see lines 14-17 and 30-34 in the WOE example in Table 3.4). At the same time, the WOEs implemented in CQT-v2 cannot mirror human WOEs move by move for a variety of reasons, including that the human tutor could point to a shared graphical representation, but the system cannot, and that the students cannot provide verbal inputs to the system.

The standard WOE for Problem 2 is shown in Table 5.6; the standard WOE for Problem 1 can be found later in this chapter, in Table 5.19 (step numbers are used for clarity, but are not included in the example when it is presented to a student).

5.3.1.2   Length and Usage of WOEs

Three other conditions in Table 5.4 concern the length and/or usage of the WOE provided to the student. The reason we explored these conditions was due to both the findings in the literature concerning adaptivity of WOEs, especially fading WOEs, and some preliminary findings by us as concerns the students from the class CS211 that we ran in Fall 2014 and Spring 2015 (the impact of the classes our subjects came from is discussed below (for other details, please refer to [175]). Specifically, we focused on the 44 students in the standard WOE condition who had used a WOE in Problem 1. First, we observed that in the Spring 2015 condition, the higher the gain, the less frequently would students bring the WOE to completion: namely, they would start the WOE, execute a few steps, but then go back to solving the problem (a self-regulated form of faded WOE, perhaps). These students may then later return to the WOE, which meant starting it all over again, but still,

Table 5.6 Standard Worked-out Example (Problem 2)

1	Right, lets look at how we can concatenate two lists!
2	We have two ascending lists, 1 to 3 and 4 to 5
3	Wouldn't it be interesting if we could make one list from 1 to 5 out of these lists. It's actually not too hard
4	All we need to do is connect the out link from the node labeled 3 to the node labeled 4
5	The tricky thing is to get the node in the first list
6	One way of doing this is to iterate from the start of the list to the end by using a temporary variable, which is Z in this case
7	We assign Z to the first node in X
8	Then use that variable to assign the next node to itself
9	and we keep doing this until we get to the end of the list
10	Now that we have access to the last node in the list X, we can hook that up to Y
11	Do a little bit of tidy up removing the extra references
12	And we're done!

they would not conclude it. The learning gain for WOE students in Spring 14 is much lower (even if not significantly so, as we will discuss below), and we observe no analogous behavior. For this reason, we also ran regressions between features of WOEs and learning gain, with pre-test as a co-variate (as in our modeling of human tutoring dialogues). We found that a binary variable *AllIncompleteWOE* which has value 1 if a student never completed the WOE in Problem 1, even if they played it multiple times, was a strong predictor: for example, for the regression that includes only pre-test and *AllIncompleteWoe*, $R^2 = 0.25$, $\beta = 0.11$, $p = 0.04$ (with $\beta = -0.48$ for pre-test).

These preliminary findings spurred us to focus on the length of time students spend on WOEs: this will depend on the length of the WOE itself, and on how the WOEs can be stepped through. Hence, we devised shorter WOEs (column ShortWOE) and imposed usage patterns on students (columns NoExit and TimeOut).

Short WOEs. Shorter versions of WOEs (as the one shown in Table 5.7) were obtained by systematically shortening the corresponding standard WOEs according to the following tenet: short examples still need to convey the same core information as the original examples but should remove non-essential steps and more verbose language.

Standard WOE steps were annotated as one of the six different types: Introduction (introduces the example); Definition (outlines the problem to be tackled); Operation (performs an operation toward the solution); Explanation (elaborates on a step); Reflection (presents an unwanted consequence of the proposed step) and Conclusion (Concluding steps of the example) – please see Table 5.19 in Section 5.3.4.1, which shows the annotation of WOE steps with

Table 5.7 Short Worked-out Example (Problem 2)

1	Here is how we concatenate two lists
2	First we need to get the last node in the first list by iteratively getting the next node of list 1 until there are no more nodes.
3	Simply hook up the last node of list 1 to the start of list 2
4	Remove extra references and we are done

their type. Short WOEs systematically eliminate Operation steps that contain a wrong step and Reflection steps which elaborate on their consequences; some standard WOEs include wrong steps that the tutor corrects right away (see steps 7 through 10 in the standard WOE for Problem 1 in Table 5.19).

The average number of steps, and of words, in standard WOEs is 11 and 167, respectively; in Short WOEs, the average number of steps is 4.5, and the average number of words, 68.

WOE Usage. As we noted above, not completing WOEs in their entirety correlated with higher learning gains; hence, we explored whether regulating the termination of WOEs may be conducive to higher gains. We looked at termination from two different angles: automatic termination of examples, and its converse, inability of students to prematurely terminate an example. Since not completing examples appears to positively correlate with higher learning, we hypothesize that automatic termination may yield higher learning gains than forcing WOE completion.

Standard WOEs in ChiQat-Tutor were enhanced to allow both of these methods of regulation to be specified on each example. While it was straightforward to set a flag stating that early termination of WOEs is not possible by disabling any button that could lead to a WOE termination (*NoExit*), automatic termination requires knowing when to terminate an example (*Time Out*). We added a time-out attribute to each standard WOE, whereby the example will terminate after a specified number of seconds. Each standard WOE has a custom time-out point that was derived from the experiment logs that we had previously collected at that point in time (namely, all experiments that included a standard WOE condition up to, and including, Fall 2015). This figure was calculated using the median of high learners' standard WOE usage durations.

5.3.1.3 Analogical Content in WOEs

Our analysis of human tutoring dialogues in Chapter 3 showed that analogy is another effective tool in a human tutor's bag of pedagogical strategies; hence, we decided to experiment with analogy within Chiqat-Tutor as well. As with implementation of other strategies in ChiQat-Tutor (or in any ITS for that matter), there are a potentially almost infinite number of choices as concerns deploying a pedagogical strategy but a very constrained number of conditions that one can run, for a whole host of practical reasons concerning experimentation. One such constraint for us was, being able to recruit a sufficient number of appropriate subjects at the right time, since we had a limited window of opportunity during which to enroll students from the classes of interest, basically within one week of them being exposed to linked lists.

In guiding our implementation of analogies, we turned back to our human–human data. We observed that for linked lists, 61% of utterances occurring in an analogy are included in a WOE (namely, 61% of utterances that occur within *begin/end* analogy markers, occur within *begin/end* WOE markers as well). Additionally, for lists, about 48% of analogies occur at the start of the tutoring session, about 32% in the middle and the remaining 20% in the final third of the session.

These two findings informed our two analogy conditions. First, in both the cases, the system will still provide WOEs, since in our data, human tutors employed both in the same session. Additionally:

1. *Initial analogy*: the student is being shown a pop-up window with an analogy for linked lists on first run. The analogy is with lining up at a movie theatre (please see Figure 5.3). This condition is motivated by the fact that, in the human tutoring data, most analogies occur at the beginning of the session. After the initial pop-up window, this condition continues as a standard WOE condition.

2. *Analogical WOEs:* in this condition, each problem has a unique on-demand analogical (long) WOE, always based on the "waiting-in-line" analogy, but possibly in different situations. For example, the analogy in Problem 1 models a line at a movie theater; the analogy for Problem 2 uses lines at check-out in a store. The length of the analogical WOE is approximately the same as the standard WOE, but the text of the WOE, and the accompanying graphical interface, model the line(s) in the situation at hand. Table 5.8 shows the analogical version of the standard WOE in Table 5.6, and Figure 5.4, the corresponding graphical interface.

5.3.2 Learning Linked Lists among Non-Majors

Before we move to analyze learning gains in our conditions, we need to address the two populations of students in our experiments, since if they differ, we should analyze their data separately. The experiments were conducted over two different courses at UIC, CS211 (Programming Practicum) and CS201 (Discrete Math and Data Structures). At the time, CS201 was a course offered for majors other than computer science (mostly for other departments in the engineering college); we considered CS201 as a valid context in which to evaluate ChiQat-Tutor, since similarly to CS211, it had Calculus 1 and an introductory class in computer programming as prerequisites. Interestingly, the CS201 class was in fact required of CS majors when ChiQat-Tutor-v1 was

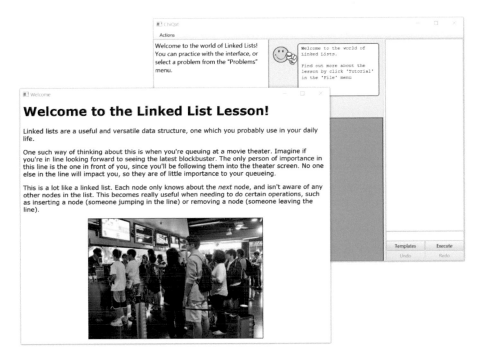

Figure 5.3 Initial Analogy (Pop-up)

Table 5.8 Analogical Worked-out Example (Problem 2)

1	Right, lets look at how we can concatenate two lists!
2	Have you ever used one of those automated checkouts at the supermarket? I find it really annoying when they break down!
3	Imagine you're at my supermarket, they have two checkouts which are REALLY busy.
4	Checkout 2, C2, is always the one that breaks. When that happens everyone from that checkout needs to go to the other checkout, C1
5	This isn't actually very hard to do, we just need to connect the last person in the first queue to the first person in the second queue
6	The tricky thing is to get the node in the first list
7	One way of doing this is to iterate from the start of the list to the end by using a temporary variable, which is Z in this case
8	We assign Z to the first node in X
9	Then use that variable to assign the next node to itself
10	and we keep doing this until we get to the end of the list
11	Now that we have access to the last person in the first queue, we can hook that up to broken checkout
12	Do a little bit of tidy up removing the extra references
13	And we're done!

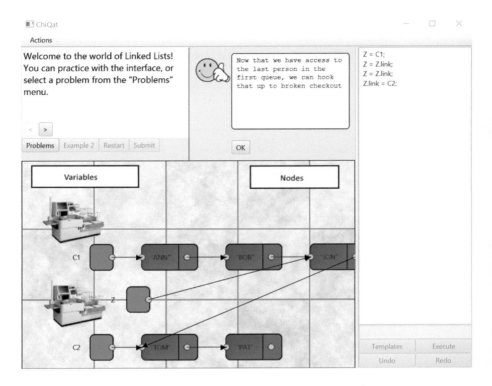

Figure 5.4 Analogical worked-out example (Problem 2).

evaluated, and the experiments run at UIC we discussed in Section 5.2 were indeed run with students in the CS201 class. In between ChiQat-Tutor-v1 and ChiQat-Tutor-v2, the Computer Science department at UIC revamped its undergraduate curriculum, and as noted, CS201 became a class offered to non majors, albeit still in engineering.

No matter the class, experiments always took place after the topic of linked lists had been introduced, and as close as possible to the relevant lectures, in general within one to two weeks. In CS201, experiments always took place in one single laboratory session; in CS211, instead, experiments always took place in four different laboratory sessions offered the same week (often the same day).

Table 5.9 includes mean pre-test, post-test and learning gain for students in the two classes. We found that the two groups of students do differ. First, the difference in pre-test score is significant, according to a t-test ($p = 0.01$), and so is the difference in learning gain ($p < 0.01$). Second, a two-factor ANOVA that predicts learning gain in terms of course and condition revealed a main effect of course ($F = 7.8194, p < 0.01$), but no effect of condition, and no interaction effect. Hence, we will start by analyzing the effects of different conditions on the students in the two classes separately. We will group them together again

Table 5.9 CS201 vs CS211 Test Scores and Learning Gains

Course	N	Pre-test		Post-test		Gain	
		μ	σ	μ	σ	μ	σ
CS201 (non majors)	70	.43	.19	.57	.22	.14	.18
CS211 (majors)	243	.50	.21	.58	.21	.08	.18

Table 5.10 CS201 - Student Distribution (Semester and Condition)

Semester	No WOE	Short WOE	Long WOE (Standard)	Total
Fall 14	16		11	27
Spring 15		5	3	8
Fall 15		17	18	35
Total	16	22	32	70

in subsequent analyses, when we investigate the initial knowledge students have, before they interact with the system.

As noted, when the evaluation of ChiQat-Tutor-v2 was conducted, the CS201 class was not part of the requirements for CS majors any longer, and was taken by other engineering, non CS majors; this may explain the difference in pre-test scores we noted earlier. Here, we focus on the experiments for CS201; those students only interacted with the NoWOE version of ChiQat-Tutor-v2 (equivalent to ChiQat-Tutor-v1.5), and then with the ShortWOE and standard LongWOE versions. For clarity, Tables 5.10 and 5.11 include only the number of subjects and learning gains of students in CS201, for the conditions they participated in. We found no significant effect of condition on learning gains. We should point out that we did find a significant difference among pre-tests among the three semesters these experiments were run, with the Spring 15 students having a significantly higher pre-test. A significant three-way ANOVA over the three semesters ($F = 4.1304, p = 0.02$) was followed by significant post-hoc Tukey tests ($p = 0.04$ and $p = 0.016$, respectively) as concerns the difference in pre-test between Spring 15 ($\mu = 0.60, \sigma = 0.26$), and Fall 14 ($\mu = 0.42, \sigma = 0.18$); and between Spring 15 and Fall 15 ($\mu = 0.39, \sigma = 0.17$). However first, the number of subjects in Spring 15 was very low.[3] Additionally, as noted, whereas we found no differences in learning gains among conditions, we found no differences among learning gains by semester either.

---

[3] Although the class was approximately 25 students, many of the them had already used the system in a previous class, on recursion. These students were therefore eliminated from analysis.

Table 5.11 CS201 - Learning Gains

Semester	No WOE	Short WOE	Long WOE (Standard)	Mean
Fall 14	.14		.24	.18
Spring 15		.15	.00	.09
Fall 15		.11	.15	.13
Mean	.14	.12	0.17	

5.3.3 Learning Linked Lists Among Majors

We will now focus on majors (students in CS211). Again, Tables 5.12 and 5.13 include only the number of subjects and learning gains of students in CS211. Students from CS211 participated in all conditions. The small number of students in the NoWOE condition in Spring 15 and Fall 15 is due to the fact that those experiments actually did not include a NoWOE condition. However, those students had previously used ChiQat-Tutor as concerns recursion (to be discussed in Section 6.2). As we will see, the recursion implementation is completely different, starting from the graphical interface, and the kind of problems that students solve (for example, only tracing, but not writing code). However, when these students used the system for recursion and created their accounts, ChiQat-Tutor still assigned them to the predefined condition active at that point (NoWOE), even if the condition was not relevant to the recursion experiment. When, one semester later, these students came back to ChiQat-Tutor, they used their previously created account, which entailed they were not assigned to one of the conditions the experiment was planned for but kept in the NoWOE condition.

TABLE 5.12 CS211 - Student Distribution (By Semester and Condition)

Semester	No WOE	Short WOE	Long WOE					Total
			Standard	No Exit	Time Out	Initial Analogy	as Analogy	
Fall 14	28		25					53
Spring 15	8		27			21	19	75
Fall 15	1	23	31					55
Spring 16				28	32			60
Total	37	23	83	28	32	21	19	243

Pre-tests do not differ among these students, neither by semester nor by condition. Similarly to the students in CS201, we did not find any differences among learning gains due to condition.

We did find one significant difference among learning gains as concerns semester of administration: after a four-way ANOVA ($F = 2.6379, p = 0.0503$) which trends toward significance, post-hoc Tukey-tests revealed that the lower

TABLE 5.13   CS211- Learning Gains (By Semester and Condition)

Semester	No WOE	Short WOE	Long WOE					Mean
			Standard	No Exit	Time Out	Initial Analogy	as Analogy	
Fall 14	.08		.00					.05
Spring 15	.16		.11			.11	.11	.11
Fall 15	.00	.13	.08					.10
Spring 16				.02	.05			.04
Mean	.10	.13	.06	.02	.05	.11	.11	

learning gain in the Spring 16 condition is trending toward significance when compared to the learning gain in the Spring 15 experiment. We will note that together, the two conditions in Spring 16 concern time constrains imposed on WOEs: this trend toward significance even if only as against Spring 15, points to the fact that imposing time usage constraints on WOEs is not conducive to learning as much as letting students pursue (and quit) WOEs at their pace, or even not using WOEs at all.

The low learning gain in the timed conditions is not even the lowest in the table: the lowest is the Standard WOE condition in Fall 14 (the "zero" learning gain in the Fall 15, NoWOE condition is due to one single student, as we explained above). Even if we found no significant interaction effects between semester and condition, this low learning gain is definitely puzzling, since it is the only group of students among the many we have run (including the 5 evaluations of ChiQat-v1) who learned nothing, and are comparable to the control condition in Table 5.1. In general, the results from that Fall 14 experimental condition should be taken with a grain of salt, and should almost be considered as a pilot. First, it was the first experiment run to evaluate ChiQat-v2, and hence, the experimenters were not as versed in providing instructions on the purpose of the study, and in motivating the subjects; second, performance in the laboratory was not linked to performance in the class, whereas in all following experiments (with both CS201 and CS211), the students were given a few points for successful completion on the activity and third, one of the four laboratory sessions that the experiment was run in was cut short by a fire alarm (in fact, we had to eliminate 4 subjects from the Fall 14 experiments who did not submit the post-test at all; they are not included in the counts and means in the tables we presented so far).

5.3.4   Learning and Initial Student Knowledge

Our analysis so far has not shown any consistent difference in learning gains across the different conditions; at the same time, so far, we have lumped all students together, independently from initial expertise. However, prior literature has shown that beginners tend to benefit more from WOEs than advanced students. [312] uses a similar concept of novice and advanced students to conclude that novice students become more time efficient with WOEs yet advanced students learn more. To explore the relationship between initial knowledge and WOEs, we will now divide students in two groups, beginners and advanced. This distinction may be important since pre-test score appears as a significant co-variate in the vast majority if not in all of our regression analyses.

First, a beginner student is defined as a student who scored in the bottom half of all students in the pre-test, while advanced students are all other students. Due to the coarse grain grading system that we had in the pretest, 34 students gained the median score of 0.46. We decided to put these students in the beginner category, in order for the groups to be more or less equal in size. In this way, the two groups, beginner and advanced, include 150 and 163 subjects, respectively. At this point, we do group students from the two original classes (CS201 and CS211) together, since the distinction between beginner and advanced is independent from course. Not surprisingly, proportionally more students from CS201 are included among beginners, as compared to CS211 (earlier in Section 5.3.2 we pointed out that the pretest of students from CS201 is significantly lower than the pre-test of students from CS211).

Tables 5.14 and 5.15 show the distribution and learning gains of beginner and advanced students, per condition. The difference in learning gains between beginners ($\mu = 0.16$, $\sigma = 0.18$) and advanced ($\mu = 0.03$, $\sigma = 0.16$) is highly significant ($t = 7.1515, df = 311, p = 6.172e^{-12}$). This result is also borne out by a two factor ANOVA: there is an effect of prior knowledge ($F = 49.5946$, $p = 1.302e^{-11}$), but no effect of condition, and no interaction. Whereas this result is not surprising, it confirms that interacting with ChiQat-Tutor is more effective for beginner students. The average learning gain for

TABLE 5.14   Beginner/Advanced Students - Distribution per Condition

Prior Knowledge	No WOE	Short WOE	Long WOE					Total
			Standard	No Exit	Time Out	Initial Analogy	as Analogy	
Beginner	25	25	56	13	13	10	8	150
Advanced	28	20	59	15	19	11	11	163
Total	53	45	115	28	32	21	19	313

TABLE 5.15   Beginner/Advanced Students – Learning Gains

Prior Knowledge	No WOE	Short WOE	Long WOE					Mean
			Standard	No Exit	Time Out	Initial Analogy	as Analogy	
Beginner	0.19	0.14	0.18	0.10	0.11	0.20	0.20	0.16
Advanced	0.04	0.10	0.01	-0.04	0.01	0.02	0.03	0.03

beginners is commensurate with the highest learning gains we had obtained with ChiQat-v1.

We further analyzed beginners and advanced students separately. For beginners, we did not find any significant difference among conditions.

For advanced students, we conducted a planned comparison ANOVA that distinguishes between learning gains as follows: we compared no WOEs ($\mu = 0.04$, $\sigma = 0.17$), short WOEs ($\mu = 0.10$, $\sigma = 0.14$), and all the variations on standard WOEs lumped together ($\mu = 0.01$, $\sigma = 0.16$). We found a significant difference ($F = 3.1384$, $p = 0.04603$). This result suggests that, when provided with worked out examples, advanced students learn more from shorter and more concise examples. Because of this result, in the next section, we will discuss how we could start personalizing ChiQAT-v2 for a specific student, based on their initial knowledge.

Even if we only found one significant difference among conditions as concerns learning gains (between short WOEs, no WOEs and standard WOEs for advanced students, as we just discussed), we will conclude by observing that for both beginner and advanced students, the lowest learning gains were obtained in the two standard WOE conditions that were constrained time-wise: *No Exit* and *Time-out* WOEs. If we were to rank-order the seven conditions according to decreasing learning gain, for both beginners and advanced, the *No Exit* condition would rank last, and the *Time Out* condition would rank second-to-last (for advanced student, the second-to-last condition is shared by the standard WOE condition). It appears that forcing students to adopt a specific WOE behavior is not productive (whereas the difference is not significant as we noted, this ranking could be seen as an instance of instant-runoff voting, also called preferential voting [201], in which bottom candidates are eliminated[4]). A caveat is that these two conditions were run in only one experiment, in Spring 2016, and were the only two conditions that were run in that setting. It is possible that some unknown features of that experiment are responsible for those results; on the other hand, as we observed earlier (Section 5.3.3), there were no differences in pre-test between the students in Spring 16, who are drawn from the CS211 class, and the pre-test of any other CS211 student group.

5.3.4.1 Mining the Logs: Predicting Initial Knowledge

As we just discussed, for students who have more initial knowledge, short examples appear to be helpful. The issue is how to gauge this initial knowledge without having to grade a pretest: this was done manually in all our experiments, which means the results of the pretest could not be used to adapt ChiQat-v2's strategies in real time. A possible way forward is to explore automatic grading, which is an active area of research. However, our data utilizes pre-test scores from tests that are not easily graded - students were required

---

[4]In instant-runoff voting, candidates are ranked by voters: the candidates who obtain the fewest first rank choices, are eliminated.

to draw images, describe issues in free text, and write blocks of code. If we continue to use such a pre-test, a sophisticated auto-grading system would be needed (indeed many automatic grading efforts are devoted to *short answer* grading [370, 442], or are proprietary [54, 348], or focus on programming per se, not on problem solving [333]). Alternatively the pre-test would have to be simplified to allow simple auto-grading. However, this would still require every user of the system to submit to a pre-test, which is not necessarily engaging for a user in less structured situations than participation in a laboratory section, and would delay them starting to actually solving problems.

Hence, we explored a second venue to attempt to gauge a student's initial ability level automatically rather than administering a pre-test before the tutorial can begin. We mined the system logs to predict how students' behavior on the first problem could predict their initial knowledge level. Because our interest in this particular instance is between whether to provide a long or a short WOE depending on whether a student is a beginner or advanced when they encounter the system, our data points of interest are the students who used Standard WOEs on the first problem: the rationale is that every student would be given a standard WOE for problem 1, and after having been classified as beginner or advanced based on their behavior on Problem 1, they would be provided with short or standard WOEs in the following problems. Namely, getting an indication of student knowledge at such an early stage would allow the system to quickly adapt its behaviour to the particular student profile.

Note that for our Machine Learning exploration, we add three more subjects to the advanced group, who were run in the very first experiment, Fall 14 (CS211). Although they were apparently assigned to the NoWOE condition, they actually accessed WOEs all the same; for this reason, we could not confidently count them in either the NoWOE or Standard WOE condition, and they were excluded from all analyses up to now. However, since we are not comparing conditions any longer, and the pre/post-test and logs of these three subjects look perfectly normal, we are including their usage of the WOE in the first problem, in the subsequent analysis.

Our logs include many facets of the student interaction: we have 88 primary log "messages", many of which are further parameterized by problem – for example, *number of positive feedback messages* or *node clicks* are collected per problem, hence there are seven different instances of these features. When we transform these log messages into features to be used for data mining, for each such feature, we also add the total across problems: for example, we compute the *total number of positive feedback messages*. This results in a total of 559 features available to mine our logs (a full description of all features can be found in Appendix E of [174]). Even for just one problem, the number of features exceeds 100: specifically for Problem 1, we are left with 127 different features. It is implausible to use an exhaustive approach and test each combination of features for the best model. We made use of *feature selection* [178, 384] to reduce the number of features to a subset of potentially important features. Different approaches to feature selection exist (e.g. wrapper and filter methods), and of these multiple different algorithms to select features. Each type of feature selection algorithm does have its advantages/disadvantages, such as some not removing redundant features. Even though reducing the feature space is desirable, it will most likely not result in the optimal set of features, and perhaps the feature selection process may remove useful features.

Therefore, we devised a bootstrapping method by getting an extended set of desirable features by use of multiple feature selection algorithms. We apply four different feature selection algorithms from the WEKA toolkit [180, 435] onto our dataset. Two of the four algorithms are based on correlations between attributes and the class (*CorrelationAttributeEval* and *CfsSubsetEval*); two others are based on Information Gain (*InfoGainAttributeEval* and *Gain-RatioAttributeEval*) – please see [174] for further details on these algorithms and their parameters. Up to the top 20 highest rated features from each selection algorithm are included in a short-list of candidate features – the short-list would therefore consist of at most 80 features. For predicting initial knowledge, this method selected 23 features.[5] Table 5.16 lists these 23 features, that we now briefly describe.

The features can be grouped as follows – about two-thirds of the features concern the standard WOE usage in Problem 1.

Problem Solving Behavior.

1. Good Submissions: how many times the student submitted a correct solution for Problem 1. Whereas this feature value is often 1, students do go back to problems and solve them again.

2. Solution attempts: the number of times a student submits a solution until their first correct solution, which is included in the count as well. For example, suppose a student submits a wrong solution for Problem 1 twice, then submits a correct solution, then goes back to Problem 1 and submits a correct solution again. The value for this feature would be 3.

Interaction with system. These 3 features encode facets of the student interaction with the system in Problem 1.

1. User acknowledgements: whenever students are given feedback, they have to click an *OK* button to proceed (see Figures 4.6 and 4.7). This feature counts the total number of those clicks.

2. Positive feedback until Success: the number of positive feedback messages until the student submits a correct solution.

3. Problem duration: the duration in seconds across all attempts at the problem.

---

[5]For other experiments we ran, if the short-list includes over 20 unique features, we further prune the list of features by removing ones that only show up in one of the algorithms. We did not do it in this case since that would have left us with only 9 features.

Table 5.16 Selected Features of Problem 1, to Predict Initial Knowledge

Type of Feature	Feature Name	Step Number	Occurrences
Problem Solving Behavior	Good Submissions	N/A	3
	Solution Attempts	N/A	2
Interaction with system	User Acknowledgements	N/A	4
	Positive Feedback until Success	N/A	3
	Problem Duration	N/A	2
WOE Features (across WOE usages)	Exit on Step [s]	7	4
	Step [s] Longest Duration	2	2
	Step [s] Longest Duration	9	1
	Step [s] Longest Duration	11	3
	Mean Step [s] Duration	1	1
	Mean Step [s] Duration	9	1
	St Dev Step Duration	N/A	1
	Total WOE Duration	N/A	1
	St Dev WOE Duration	N/A	1
WOE Features (first WOE usage)	Step [s] Duration	1	1
	Step [s] Duration	2	1
	Step [s] Duration	6	1
	Step [s] Duration	8	1
	Step [s] Duration	9	1
	Step [s] Duration	11	1
	Mean Intro Step Duration	N/A	1
	Mean Explanation Step Duration	N/A	3
	Mean Reflection Step Duration	N/A	1

WOE features (across usages). Very often, students start the WOE, exit it and come back to it. These features are computed across all WOE usages in Problem 1.

1. Exit on step *[s]*: how many times the WOE has been exited on step *s*. This feature is parameterized for each step (the WOE in Problem 1 has 14 steps, see Table 5.19). The feature selection algorithms retained only the count of exits for step 7.

2. Step *[s]* longest duration: across the usages of the WOE, what is the longest time the student spent on step *s*? Again, this feature is parameterized for each step. This time, the feature selection algorithms retain this feature for three steps, 2, 9 and 11.

3. Mean step *[s]* duration: the mean duration of step *s* across WOE usages. Again, this feature is parameterized by step *s*. In this case, the feature selection algorithms retain this feature for steps 1 and 9.

4. St Dev Step Duration: standard deviation for total step durations (across all steps) during all WOE usage

5. Total WOE duration: the sum of the times spent across all usages of the WOE in Problem 1

6. St Dev WOE Duration: standard deviation of WOE usage durations for Problem 1

WOE features from the first usage of the WOE. As we noted, students can start a WOE, exit and come back to it as many times as they want. This group of features concerns the first time they used the WOE in problem 1.

1. Step *[s]* duration: the duration of step *s* on first usage of the WOE. Again, this feature is parameterized by step *s*. In this case, the feature selection algorithms retain this feature for steps 1, 2, 6, 8, 9, and 11.

2. Mean *[type]* step duration: the mean duration for steps of a certain type during the first usage of the WOE. Recall that standard WOE steps were annotated for one of six types. This feature is parameterized by step type: three types are retained, intro(duction), explanation and reflection. The WOE in Problem 1, with steps annotated for type, is repeated in Table 5.19. Since there is only one reflection step, Step 8, the corresponding feature is basically equivalent to the time spent on Step 8.

Having focused on these 23 features, we ran exhaustive experiments with a battery of 16 ML algorithms from the WEKA toolkit [180, 435]. These 16 algorithms were selected from a list of 49 potential algorithms, many of which were eliminated due to them generating invalid models given the training dataset (such as the model generation process throwing an error on creation). The 16 algorithms we ran are listed in Table 5.17, by type. Here, we briefly describe the various classes, please see [174] for details on both how the 16 algorithms were chosen, and the parameter settings required by many of the 16 algorithms.

Naive Bayes. This class includes algorithms at whose core is Bayesian conditional dependency between the class of interest and features; the "naivete"' comes from considering features independent from each other. The variations on Naive Bayes we ran include the standard algorithm, the multinomial version that models each individual conditional probability via the multinomial distribution, and the updateable version,

Table 5.17 Machine Learning Algorithms

Class of Algorithm	Specific type
Naive Bayes	Standard
	Multinomial
	Updateable
Regression	Logistic
	Simple Logistics
	ClassificationViaRegression
	Stochastic Gradient descent (SGD)
Neural Networks	Multilayer Perceptron
	Voted Perceptron
Support Vector Machine	
Decision trees	J48
Rules	DecisionTable
	One Rule
	InputMappedClassifier
Ensemble Classifiers	AdaBoostM1
	RandomForest

which updates its model one datapoint at a time (as opposed to processing all the data in batch mode). Naive Bayes algorithms are often used as baselines in Machine Learning experiments.

Regression. The algorithms in this class are all based on regression. They include Logistics Regression; Simple Logistics which has built-in attribute selection; SGD as applied to binary logistics regression.

Neural Networks. In this era of AI success due to deep learning [258], it is important to experiment with some of those formalisms, namely, neural networks with several layers of nodes. In our case, we employed the multilayer perceptron (MLP), a class of feedforward neural network, with at least three layers. The second algorithms is the Voted Perceptron [135], which uses multiple weighted perceptrons. Every time an example is wrongly classified, a new perceptron is initialized with the final weights of the last perceptron. Each perceptron will also be assigned an additional weight that corresponds to how many examples it correctly classifies before wrongly classifying one; the output will be a weighted vote of all perceptrons in the model.

Support Vector Machine (SVM). We only ran one algorithm in this case. SVMs return hyperplanes separating classes of interest [89].

Decision Trees. J48 is a classic implementation for decision trees [345]. A decision tree is a structure in which each internal node represents a test on an attribute, each branch represents the outcome of the test and each leaf node represents a class label.

Decision Rules. The Decision Table algorithm learns decision tables, namely, pairings between conditions and actions; such pairings can be interpreted as rules. Other algorithms directly learn rules. The ones we used are: OneR, short for "One Rule", generates one rule for each predictor in the data, then selects the rule with the smallest total error as its "one rule". Input Mapper is not a decision rule algorithm per se, but rather, a wrapper that *addresses incompatible training and test data by building a mapping between the training data that a classifier has been built with and the incoming test instances' structure* [435]. However, we used it with its default classifier, ZeroR, which derives the simplest possible rule, the one that defines the majority class.

Ensemble Classifiers. Ensemble algorithms combine the predictions from multiple models. We employed three of them: AdaBoost and Random Forest, which are applied to decision tree learning, and Classification via Regression. Specifically, AdaBoost builds an ensemble of short decision tree models, each with a single decision point (referred to as *decision stumps*). In the first iteration, trees are built as usual; the algorithm then continues to train additional models until the desired accuracy is obtained or no further improvements are possible. Each model is weighed based on its performance; predictions from all of the models on new data are weighed accordingly, to choose the final classification. Random Forest builds a multitude of decision trees at training time and outputs the class that is the mode of the classes (classification) or mean prediction (regression) of the individual trees; in our case, we used classification. As concerns Classification via Regression, the class is binarized and one regression model is built for each class value. The different models are then appropriately combined.

We ran exhaustive experiments with all the 16 algorithms, with all possible subsets of features out of the 23 above, up to a cardinality of 6. Namely, for each algorithm, we ran almost 150,000 experiments (according to the formula $\sum_1^6 \binom{n}{23}$), with each subset of the 23 features such that $1 \leq s \leq 6$, where s is the size of the subset. Two algorithms were consistently among the top 10, J48 (decision trees) and MLP (multilayer perceptron). Table 5.18 reports the results and the features these top ten algorithms used. Note that all results are the averages of the respective metrics (Accuracy, Precision, Recall, F-score) computed on the individual folds during cross-validation.

Table 5.18 highlights that features related to problem-solving behavior and to interaction with the system are quite predictive of the novice/advanced initial knowledge state: both problem-solving behavior features are selected by over half of the ten top models, and the *Positive feedback until Success* feature (i.e., how much positive feedback they received until they submitted a correct solution) is included by all but one of the top models. Several WOE features are also included: to illustrate them, we will refer to the actual steps

Table 5.18 Top Models for Predicting Initial Knowledge

Feature Type	Feature	J48	J48	J48	J48	MLP	J48	J48	MLP	MLP	J48	J48	Frequency
Problem Solving Behavior	Good Submissions	○	○			○	○	○	○		○		6
	Solution Attempts	○	○	○	○	○	○	○	○		○	○	8
Interaction with System	User Ack.	○				○		○					1
	Pos. Feedback Until Success	○	○	○		○	○	○	○	○	○	○	9
	Problem Dur	○											1
WOE Features (across usages)	Exit on Step 7				○	○	○	○	○		○		5
	Step 2 Longest Dur.			○					○		○		3
	Step 9 Longest Dur.						○		○				2
	Step 11 Longest Dur.						○						1
	Mean Step 9 Dur.					○							1
	Mean Step 1 Dur.												0
	StDev Step Dur.				○					○			1
	Total WOE Dur.												1
	St Dev WOE Dur.			○		○							1
WOE Features (first usage)	Step 1 Dur.	○					○				○		2
	Step 2 Dur.		○				○						1
	Step 6 Dur.				○	○		○	○				4
	Step 8 Dur.												1
	Step 9 Dur.			○		○		○	○				4
	Step 11 Dur.												1
	Mean Intro Step Dur.						○						1
	Mean Explanation Step Dur.					○			○	○			2
	Mean Reflection Step Dur.												0
	Total Features used	4	6	6	6	6	6	5	6	6	6	5	
	Accuracy	.765	.765	.765	.765	.765	.757	.757	.757	.757	.748	.748	
	Precision	.765	.765	.766	.768	.768	.757	.759	.767	.767	.748	.749	
	Recall	.765	.765	.765	.765	.765	.757	.757	.757	.757	.748	.748	
	F-Score	.765	.765	.765	.764	.764	.756	.756	.753	.753	.748	.748	

Table 5.19 Standard Worked-out Example for Problem 1 (Step types are included)

1	Introduction	OK, (USERNAME), we're going to take a look at how we can insert items into a linked list
2	Introduction	Take a look at this list
3	Definition	What we want to do is to insert a '3' in between '2' and '4'
4	Operation	The very first step is simple, let's create a new node with a value of '3', and why don't we call it 'Z'
5	Explanation	Now we need to insert a node AFTER '2', the one that's flashing
6	Operation	Firstly, we should get access to the second node. This can be done by going through the root, T, and getting its next node. This can be assigned to a variable, S
7	Operation	From here, we could assign 2's next pointer to the node containing 3, like so
8	Reflection	However, there is a problem here, think about it...
9	Explanation	How do you reattach the node containing 4? The connection now has been lost!
10	Explanation	Let's take a step back and see how we can do this without losing this vital connection
11	Operation	Let's connect the node containing 3 to the node containing 4
12	Operation	Now let's do what we done before and connect 2 to 3
13	Operation	and then tidy up some of the references, this being S and Z
14	Conclusion	And there we go, 3 has been inserted into the list between 2 and 4!

in the standard WOE students use, included here again for convenience in Table 5.19.

The following steps figure prominently in the top ten models (they are chosen by more than one model), even if through different features: Step 1, Step 2. Step 6, Step 7 and Step 9. Steps 1, 2 and 6 affect the final models through their duration, in the first WOE usage (steps 1 and 6), or across all WOE usages (Step 2). Steps 1 and 2 introduce the example; note that they appear once more, together, through the feature *Mean Intro Step Dur*. Step 6 is a rather demanding Operation step, since it compactly describes a conceptually complex step. However, the time it takes students to read it is also affected by its length: Step 6 is much longer than the other steps (the average length of a step in this example is $\mu = 15$ words, $\sigma = 6$: Step 6 is 32 words long). More revealing are the roles played by Steps 7 and 9. First, the important characteristic of Step 7 is the count of how many times a student leaves the WOE on this step. Step 7 demarcates where in this standard WOE a potential solution is presented, but its negative side-effects may not have been revealed yet. Step 9, an explanation step which points out the problem with the proposed solution, appears seven times in the top ten models, either as concerns its longest duration or its mean duration across all usages of the WOE in the first problem, or as it concerns its duration in the first usage of the WOE.

Figure 5.5 includes the two top models, both decision trees. The class to be inferred is indicated as 0 (novice) or 1 (advanced). Leaves indicate the classification that is returned if that path in the tree is followed: for example, according to the first tree, a student is classified in the advanced category, if they had at least one good submission, received more than 14 positive feedback messages before succeeding, spent less than 4 seconds and 814 ms on step 6, and submitted at least two incorrect solutions before submitting a correct one. The number in parenthesis refers to the datapoints that the specific leaf classifies in total, with the second number indicating how many of those are incorrect. Following the path just described, 10 students are covered by that leaf, of which one, incorrectly.

The second tree has the same structure at the root and first level as the first tree, based on number of *Good Submissions* and of *Positive Feedback until Success*. After that, the second tree focuses on the duration of Step 9 during first usage, and then on the interplay between number of *Solution Attempts*, and very specific distinctions on the number of *Positive Feedback until Success (PFuS)*: in the branch rooted in *SolutionAttempts=1*, we are considering two intervals for PFuS, $14 \leq PFuS \leq 19$ and $PFuS > 19$.

It is interesting to note that in both the trees, when the tree splits on number of *Solution Attempts*, it is often the case that the higher this number, the more advanced the student is (please see branches `Solution Attempts > 2` and `Solution Attempts > 6` in both trees). Recall that this feature captures the number of times a student submits a solution until their first correct solution, which is included in the count as well. A priori, one would expect that advanced students have fewer *Solution Attempts*, in that they need fewer incorrect attempts to reach the correct solution. A possible explanation for this counterintuitive result is that more advanced students realize they are on the wrong path more promptly than novice students, and that they prefer to start from the beginning (which they do by submitting an incorrect solution) rather than continue floundering; or alternatively, that advanced students explore the space of solutions by obtaining feedback that a solution is incorrect, and prune out incorrect solutions in this fashion. We attempted to investigate these hypotheses, for example, by checking for an inverse correlation between number of *Solution Attempts* and number of *Undo* operations, but we did not manage to obtain additional insight into advanced students' behaviors.

5.3.5 Chiqat-Tutor, Version 2: Summary of Findings

With ChiQat-Tutor-v2, we ran a large number of conditions over two academic years. The conditions differed in whether the students were majors or not; whether they had access to WOEs, and if yes, in which form: short, or if

Top tree:

```
Good Submissions = 0: 0 (23.0/5.0)
Good Submissions > 0
|    PosFeedback Until Success <= 14: 1 (48.0/11.0)
|    PosFeedback Until Success > 14
|    |    Step 6 Dur. (First Usage) <= 4" 814ms
|    |    |    Solution Attempts <= 2: 0 (17.0/6.0)
|    |    |    Solution Attempts > 2: 1 (10.0/1.0)
|    |    Step 6 Dur. (First Usage) > 4" 814 ms
|    |    |    Solution Attempts <= 6: 0 (14.0)
|    |    |    Solution Attempts > 6: 1 (3.0/1.0)
```

Second best tree:

```
Good Submissions = 0: 0 (23.0/5.0)
GoodSubmissions > 0
|    PosFeedback Until Success <= 14: 1 (48.0/11.0)
|    PosFeedback Until Success > 14
|    |    Step 9 Dur. (First Usage) <= 2" 049ms
|    |    |    Solution Attempts <= 2
|    |    |    |    Solution Attempts = 1
|    |    |    |    |    PosFeedback Until Success  <= 19
|    |    |    |    |    |    PosFeedback Until Success  <= 16
|    |    |    |    |    |    |    Problem Duration <= 7'33": 1 (2.0)
|    |    |    |    |    |    |    Problem Duration > 7'33": 0 (2.0)
|    |    |    |    |    |    PosFeedback Until Success > 16: 1 (2.0)
|    |    |    |    |    PosFeedback Until Success > 19: 0 (3.0)
|    |    |    |    Solution Attempts  =2: 0 (8.0/2.0)
|    |    |    Solution Attempts > 2: 1 (11.0/2.0)
|    |    Step 9 Dur. (First Usage) > 2" 049ms
|    |    |    Solution Attempts  <= 6: 0 (13.0)
|    |    |    Solution Attempts  > 6: 1 (3.0/1.0)
```

Figure 5.5 The top two decision trees to distinguish between novice (0) or advanced (1) students (PosFeedback = Positive feedback).

long, whether the WOE had an analogical flavor; or whether time constraints were imposed on WOE execution. We did not find as many significant results as we would have liked. What we established is that non-majors and majors did differ both in pre-test and in learning gains, suggesting that, no matter the condition, a system like ChiQat-Tutor-v2 is helpful to those students who may have lower knowledge. In fact, the average learning gain for non-majors interacting with ChiQat-Tutor-v2 is $\mu = 0.14$, identical to that with human tutors (and to the learning gain with ChiQat-Tutor-v1.5, as well).

As concerns the different strategies that students were exposed to in the different versions of ChiQat-Tutor-v2, we did not find significant differences. The only consistent finding, which was borne out both analyzing all students, and distinguishing beginner versus advanced students, is that learning gains in time constrained conditions are the lowest, even if they are not significantly different from learning gains in other conditions (recall that there are two such conditions, *NoExit* in which the student must step through the WOE till completion, and *TimeOut*, where a time limit is imposed on each WOE, so that the student will not be able to complete it).

Once we started analyzing beginner versus advanced students, we found a trend toward significance, as concerns advanced students learning more from short WOEs than from long WOEs. Given this, we embarked on mining the system logs in order to be able to distinguish beginners from advanced students; if we knew this, we could provide them with short or long WOEs, according to their knowledge level. Our assumption is that for ChiQat-Tutor-v2 to be more usable *in the wild*, such determination should not depend on administering a pre-test, that to be meaningful needs to be complex enough, and hence, would need manual grading. One window into students' prior knowledge is their behavior on the first problem in the system, including their usage of the associated WOE. We ran an exhaustive suite of Machine Learning algorithms on the logs of all students who worked on the long WOE in Problem 1. The best models perform at an accuracy of 0.765 (the baseline is 0.526), as concerns distinguishing beginner from advanced students. Additionally, the best models highlight the predictive power of features related to the number of correct submissions, the role of positive feedback (which we had already ascertained in ChiQat-Tutor-v1.3 through ChiQat-Tutor-v1.5), and the time spent on certain specific steps in the WOE for Problem 1.

III

Extending ChiQat-Tutor

CHAPTER 6

Beyond Linked Lists: Binary Search Trees and Recursion

With Mehrdad Alizadeh

Lexis Nexis

With Omar AlZoubi

Jordan University of Science and Technology

CONTENTS

IN Chapters 4 and 5, we focused on many different aspects of ChiQat-Tutor, but our main modeling and experimentation were centered on one data structure, linked lists. In the course of our work, we also addressed other important topics for introductory CS education, specifically Binary Search Trees and recursion. We devote this chapter to those models. This chapter provides a high-level demonstration of how to extend ChiQat-Tutor to different topics.

The next chapter will provide extensive information for practitioners, and a tutorial, on how to do so in practice.

6.1 BINARY SEARCH TREES

Binary Search Trees, or BSTs, are one of the simplest two-dimensional structures, while lists and stacks are one-dimensional data structures; hence, BSTs are often the first two-dimensional data structure students are exposed to. BSTs are often used in applications since, as their name implies, they are optimized for search on ordered sets: each node contains a value larger than any values contained in its left subtree (i.e., the tree whose root is the left child of the node in question), and smaller than any values in the right subtree (i.e., the tree whose root is the right child of the node in question). For example, in the BST in Figure 6.1, the root of the top tree contains 8, which is larger than any values contained in the nodes to its left (its whole left subtree), and smaller than the values contained in any node to its right, in this case, only 9. The same applies if we descend lower in the tree.

BSTs had been included in our human–human tutoring data, as we described in Chapter 3. The BST module we developed is similar to the linked list tutorial, and gives students the ability to search and manipulate BST structures. The challenge for students is to manipulate the BST in such a way as to maintain its ordering property.

The BST curriculum includes 6 problems: one on insertion; two on deletion; one on finding the minimum node in the BST (via iteration); one on finding the node with value X (via iteration) and one on finding the node with value Y (via recursion: however note that the recursion model for this specific problem was set up ad-hoc, and it is not related to the recursion architecture that we will discuss in the next section).

A student starts a problem with an initial model, as shown in Figure 6.1 for Problem 4 (note that Problem 4 presents two different trees to make sure the student solves the problem in a general way). The student solves the problem by writing code in the rightmost window, and the interface updates the model according to the execution of the student's code, as in all other versions of ChiQat-Tutor. Likewise, feedback is provided, but only syntax, execution and final, neither reactive nor proactive. To generate reactive and proactive feedback, the PKM is required; in turn, to train the PKM, a sizable amount of interaction logs is needed, as we discussed in Section 4.5. The BST version of ChiQat-Tutor never moved beyond piloting, namely, we never used it with an adequate number of students.

Hence, feedback wise, ChiQat-Tutor for BSTs is equivalent to CQT-v1.2 (please see Table 4.1 in Section 4.4). For example, Figure 6.2 shows what happens if the student tries to execute the snippet of code shown in the feedback window. The error message is contained in the feedback window itself that the student would scroll through, but for the purpose of presentation, here, it is shown on the right hand side of the figure.

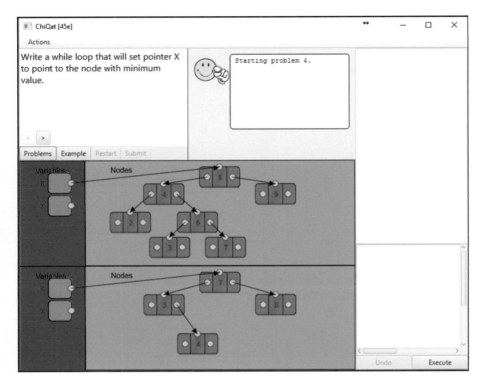

Figure 6.1 The beginning of Problem 4 for BSTs

When the student submits his/her solution, the solution evaluator evaluates a number of constraints. For example, a constraint concerns whether the model represents a BST: the evaluator checks that a parent node value is lower than the value in the left child and higher than the value in the right child. If a constraint fails, then the feedback manager posts a relevant message; otherwise, a congratulation message appears: "Good job! You have solved the problem".

6.1.1 Pilot Evaluation

The BST version of ChiQat-Tutor was formally evaluated via a small pilot run independently from a specific class curriculum, unlike the numerous evaluations we ran on linked lists, and on recursion. Twelve subjects, eight during Fall and four during the subsequent Spring were recruited from the CS251 course at UIC, a required course that focuses on data structures and is most often taken during sophomore year: this is where students would be exposed to BSTs for the first time. Students came to our research laboratory; after informed consent, they took a pre-test consisting of three problems, used the BST module in ChiQat-Tutor and took the same post-test afterwards.

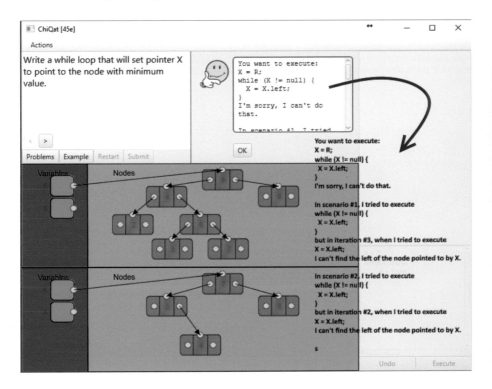

Figure 6.2 An Error during Problem 4 for BSTs.

On average, these 12 students attempted 5 problems, and submitted 4.17 correct solutions, and 3.25 incorrect solutions each (recall that after a student submits an incorrect solution, s/he can attempt to solve the same problem again). They showed a small average learning gain $\mu = 0.07$ (the cumulative test score is normalized to 1, as usual). If we eliminate 4 subjects who were almost at ceiling ($\mu \geq 0.9$ on pretest), the average learning gain is slightly higher, $\mu = 0.09$.

This pilot evaluation was useful to show that the BST module was robust enough to be formally tested, that the problems were appropriate for the students, and that the students could successfully advance through the curriculum. It is clear that a more extensive evaluation, linked to a class curriculum, would be necessary to truly assess the effectiveness of this module. Additionally, since algorithms on BSTs and more complex types of trees are essentially recursive, the BST module should be more closely aligned with the Recursion version of ChiQat-Tutor, to which we turn next.

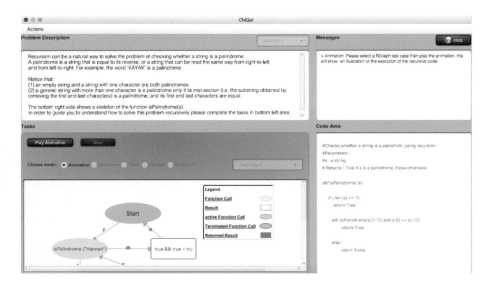

Figure 6.3 Recursion Interface in ChiQat-Tutor. Top left: a problem and its explanation. Top right: description of task (animation). Bottom left: current task, task list and recursion graph. Right: recursive code.

6.2 RECURSION

Recursion is a fundamental concept in computer science: it is a mathematical concept, a way of formalizing algorithms, and a programming technique [287]. At the same time, it is a powerful computational approach that splits a problem into subproblems of the same type, until what is called the base case is reached; at that point, the results of the smaller problems are back-propagated and composed in some appropriate fashion to compute the answer to the larger problem(s).

An example of a recursive function to compute whether a string of characters is a palindrome (reads the same backward and forward) is presented in the right pane of Figure 6.3 (since to make the figure readable the bottom left window containing the recursion graph (RGraph) is cut, the whole RGraph is shown in Figure 6.4 - we will discuss RGraphs at length shortly). This problem can be solved by checking whether the first and last characters of the string are the same, eliminating them and recursively applying the same function until we are left with the empty string, or with one character only (base case), which by definition are palindrome.

Students struggle to learn this type of problem decomposition [95]; indeed, computer science pedagogy maintains that recursion is intrinsically difficult to master [339], and to teach [140]. The reasons for this are varied.

According to [406], students sometimes have difficulty recognizing different invocations of the same function; since in a programming environment, each

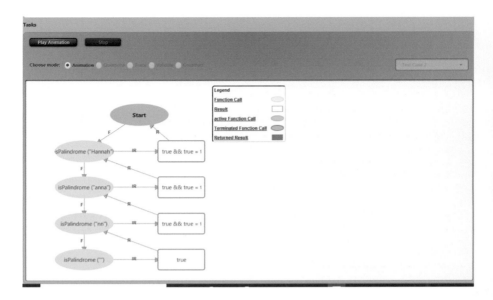

Figure 6.4 Full recursion graph for the *Play Animation* task from Figure 6.3.

recursive call is suspended until it can be executed, students struggle with the visualization of the program execution, with the bookkeeping required by each call, and with the backward flow of control after reaching the base case. More in general, difficulties come from recursion being a less familiar concept to students than iteration, and the lack of everyday analogies for recursion. A compounding reason is that recursion is traditionally taught after iteration, namely, after students have built up preconceptions about self-referential processes based on their experience with looping [411].

[181] presents a recent detailed inventory of student difficulties with and misconceptions about recursion; the authors extracted this inventory from a large corpus of approximately 8000 answers to recursion questions from examinations given to students over three semesters in a traditional CS2 course. These findings cover a variety of issues pertaining to backward flow, recursive calls and the base case. Interestingly, in one such misconception, students assume there is a global variable that is updated by each recursive call; it appears to be related to the "Loop" conception of recursion that we will discuss below, and that was studied 35 years earlier in [218].

6.2.1 Models for Teaching Recursion

There are two main approaches to teaching recursion, conceptual models and visual aids.

Conceptual models for teaching recursion help learners construct correct mental models of the concept [164, 373, 436]: these models include mathematical induction, abstract and conceptual discussions of recursion, process tracing and structure templates of recursive code [436]. The rationale is that correct mental models of a process will allow the learner to predict the behavior of that process, albeit not necessarily always correctly [218].

On the other hand, other CS educators believe that in order to conceptualize recursion and to be able to utilize it in writing code, the learner must be able to visualize the problem and in particular how solutions to smaller problems of the same type are merged to solve the original problem [43, 406]. Hence, the visualization approach relies on algorithm animation, and on a variety of program visualizations, including videogames.

6.2.1.1 Conceptual Models

Initially, researchers proposed the mathematical induction model [127], in which recurrence relations are used to teach recursion to students [430]. Recently, Lewis [261] examined the role of four modes of algebraic substitution techniques on students learning to trace linear recursion, under the hypothesis that differences between these techniques might help students understand the recursion process as it executes on computers. However, beginning computer science students might not possess the necessary level of mathematical skills that would allow them to take advantage of algebraic substitution techniques in order to understand recursion [127, 430].

Other researchers stress a declarative, abstract level of problem decomposition and advocate applying a divide-and-conquer strategy at the problem level, independently of any programming implementation. [158] found that students who utilized this model were significantly better at formulating recursive programs as compared to those taught the process of recursion execution. [436] examined five different conceptual models, of which three concrete and two abstract. The three concrete ones were "Russian Dolls" (whose meaning is precisely what its name describes); "Process Tracing" and "Stack Simulation" (that refers to the system stack mechanism of process execution). The two abstract ones included the "Mathematical Induction" that we already discussed; and "Structure Template", where samples of recursive problems are provided with explations (in the spirit of worked-out examples). [436] found that students learned more with the concrete models; however, when controlling for type of learner (concrete vs abstract), the abstract models were better for abstract learners.

[218] identified a number of mental models of recursion possessed by novice programmers, as opposed to the correct "Copies" model, in which each recursive invocation generates a new instantiation of the procedure. The most common wrong model is the so-called "Loop" model, in which the recursive procedure execution is viewed as a single entity instead of a series of new copies. The parameters to the recursive procedures are seen as boxes which hold values which are displaced by subsequent values. As previously noted, other researchers had suggested that recursion should be tauught before loops

and not vice-versa [411]. At the same time, [218] cautioned that having acquired the Copies mental model is not sufficient to determine what students really know about recursion, since students can make predictions about recursive procedures behavior without fully understanding recursion. [430] built on [218] and evaluated a number of conceptual models to teaching recursion, including mathematical models and the Copies model. They found that the Copies model is more viable than the other models.

Other older research used structure templates and worked-out examples for teaching recursion in the context of LISP programming [339]. Students were guided in their early attempts by abstract representations of the structure of recursive examples, as common in approaches that use worked-out examples (see Sections 2.1.2.1 and 2.4.2.2). However, [357] pointed out that structured templates and examples should not be used too early, since this can result in students adopting the syntactic model, where novices make predictions based on the position of the different program segments (e.g. base case) in the program layout itself.

6.2.1.2 Program Visualization

In the last 20 years, probably also due to the improvements in graphics and animation, program visualization has often been suggested as an important tool to support teaching of recursion, and in fact, of algorithms in general. For example, [95] experimented with Alice, a well known 3-D interactive graphics programming environment [82]. Alice has an object oriented flavor and it offers full scripting environment for 3-D object behavior (e.g., animals and vehicles) in a virtual world. Students can control object appearance and behavior by writing simple scripts, and in so doing, gain insight into how recursion unfolds. However, [109] cites two limitations of the Alice system which in particular affect visualization for recursion: a) Alice does not allow to create new objects programmatically and b) Alice does not allow the inspection and modification of functions from within an executing program.

[410] used VisualCode, a visual notation that uses colored expressions and a graphical environment to describe the execution of Scheme programs. Students in the experimental group (who were instructed with VisualCode) outperformed those in the control group in both evaluation and writing of recursive programs.

Similarly, [406] used the Cargo-Bot video game for introducing recursion to novice programmers. In this game, players control virtual robots by creating programs using a simple visual language; the aim is to move a set of crates to a specified goal configuration via a robotic arm. One of the interesting features of Cargo-Bot is that it supports recursion but not looping. The authors measured whether students who played Cargo-Bot were able to transfer their experiences in playing Cargo-Bot to solve recursive problems in Java. Results from a controlled experiment showed significant improvements in students' understanding of recursion using this technique.

Animation tools that show algorithm executions are potentially very useful to teach recursion, since animation appears to make a challenging algorithm more understandable, thus positively affecting learning [225]. The question is how students should use the animations that the teacher prepared: in order of increasing engagement, simply watch them, run them by providing different inputs or manipulate them, as argued by [43] and [393]. It appears that there is no consensus on the benefits of algorithm animation as a learning aid. [208] provides a meta-analysis of 24 studies exploring the effects of algorithm visualization (AV) on learning algorithms, and shows that the important variable is how students use AV rather than what AV shows them. In Section 6.2.4, we will discuss a small but negative effect of animations on learning in our experiments.

Visual aids specifically targeting recursion were proposed by [369], and are called Recursion Graphs (RGraphs) (we adopted them in ChiQat-Tutor; the one from the bottom left pane of Figure 6.3 is presented in its entirety in Figure 6.4). RGraphs present the activation sequence of recursive functions; with respect to recursion trees, they provide a detailed representation of the sequence of recursive calls including intermediate results for each such call. An RGraph is a directed graph with two sets of nodes: one for recursive calls (indicated by ellipses) and one for pre/post processing statements of those calls (indicated by rectangles). The RGraph is built level by level from top to bottom (i.e. breadth-first) with directed edges which denote the activation sequence. RGraphs are traceable since they present the detailed invocation sequence from one level to the next. The proponents of RGraphs [369] showed that RGraphs positively affected students' learning of recursion.

Before moving to our own approach, we would like to circle back to [181]. Rather than subscribing to a specific conceptual or visualization model of recursion, the authors developed an interactive tutorial, RecurTutor, that plays the role of an e-book. It includes 10 example presentations that specifically address the inventory of difficulties and misconceptions they uncovered from their analysis of students answers to recursion questions, that we mentioned earlier, and importantly, 20 tracing exercises and 20 programming exercises. For the exercises, feedback is provided as concerns correctness of the answer. The authors found significant differences in learning between students who used RecurTutor, and those who did not, on a variety of different problems on recursion (tracing, detecting infinite recursion, and writing recursive code). The effect was moderate. The authors were unable to assess which specific components of RecurTutor affected learning; we will address the specific tasks that appear more conducive to learning in ChiQat-Tutor, when discussing our results in Section 6.2.4.

6.2.2   A Hybrid Model for Teaching Recursion in ChiQat-Tutor

In developing our own recursive model within Chiqat-Tutor, we adopted a hybrid approach, that combines the respective strengths of conceptual models and visual aids, and at the same time, mitigates the misconceptions students may develop when using only one of the two. ChiQat-Tutor uses code structure templates, which originate in the abstract model of teaching recursion, and takes advantage of the visual model of RGraphs.

As concerns using structure templates of recursive code and worked-out examples (WOEs), we extensively discussed the cognitive import of worked-out examples and our own results on using WOEs for linked lists in Sections 2.1.2.1 and 5.3.1 respectively. In the recursion module of ChiQat-Tutor, worked-out examples are used organically, via say the "play animation" task (see below), not via a button that the student has a choice to click in order to obtain an example, as in ChiQat-Tutor-v2.

ChiQat-Tutor's modular architecture was ideally suited to support the usage of RGraphs. We extended the use of RGraphs in our system by implementing several interactive tasks described in Table 6.1, and illustrated in the interface shown in Figures 6.3 and 6.4 . Students can interact with RGraph representations of different recursive problems (e.g., factorial, palindrome). The RGraph-based interactive tasks are progressively more complex, and can help students identify the recursive structure of problems, and the features of recursive solutions to these problems. These tasks integrate features of some of the different conceptual models to teaching recursion that we discussed earlier. For instance, the stack simulation approach motivates the animation task, whereas the Copies mental models comes across in different guises via the tracing, validating and constructing tasks; performing these tasks will provide students with different inroads into the expert Copies model.

It is also important to mention that many operations on linked lists and BSTs are naturally performed with recursive solutions. Thus, improving students understanding of recursion will allow them to apply recursive solutions to these types of data structures. We plan to add tasks that allow students to manipulate linked lists and BSTs using recursion. These tasks include traversal of the structure, and addition and deletion of nodes. ChiQat-Tutor will serve as an effective visualization tool with which hidden effects of nested function calls would become evident when applied to a whole range of problems that can be solved recursively.

6.2.3   Evaluation of the Recursion Module

We evaluated the recursion module with multiple experiments conducted both at UIC and at CMUQ. Table 6.2 provides a summary. We will first describe the experiments in general; second, discuss the CMUQ and UIC experiments seperately since they differ along a number of dimensions; but third, bring the two groups of students together as concerns the correlations between tasks performed with the system and learning.

Table 6.1 Task Description

Task Name	Description
Answer	Answer multiple choice questions on the current recursive problem.
Play animation	Play prepared animation and observe execution order of recursive calls. Nodes' color meaning: green indicates an active function call; grey a terminated call or an intermediate result.
Trace	Click on the RGraph nodes and follow right order of execution. Nodes' color will change as users make progress.
Validate	Given sample snippet of code, two types: a) Incomplete RGraph: fill the partial RGraph, and validate the solution. b) Incorrect RGraph: correct errors, and validate solution.
Construct	Build RGraph for given recursive code snippet. The first few nodes of the RGraph are provided. Validate solution after finishing constructing the RGraph.

6.2.3.1 Experimental Protocol

Students started by taking a pre-test (limited to 10' maximum). The pre-test comprised two problems and related subproblems; in total, students had to answer 5 questions, two of which had the students make use of RGraphs in order to illustrate recursive decomposition. Then, depending on condition, students were asked to either solve problems on paper or to work with ChiQat-Tutor; either condition was capped at 30 minutes (note that students at CMUQ only worked with the system). The same three recursion problems were provided in the two conditions: *Power-of-two, Factorial, and is-Palyndrome*. For each problem, students had to perform all of the tasks presented in Table 6.1. Those tasks are ordered, at least partially, by the degree of cognitive load that they require of the student; *Answer questions* is unrelated to using the RGraph per se.

One difference between the paper and system conditions is that animation is the one task from Table 6.1 that is not doable on paper; otherwise, all the other tasks (trace, validate correct and incorrect RGs, and construct RGs) were conducted on paper as well. Although students were asked to finish all three recursion problems, completing any of the problems, or of the related tasks, was not enforced. Additionally, no fixed order was required as concerns problems and tasks, namely, students were free to look at them in any order they preferred, in both paper and system conditions. After the 30' devoted to problem solving, students were asked to answer the same questions from the pre-test, with the same 10' time limit (namely, the pre- and post-tests were identical).

Table 6.2 Recursion Experiment Student Distribution (by Semester and Condition)

School	Semester	Paper	System	Total
CMUQ	Fall 14	0	16	16
	Fall 15	0	21	21
UIC	Fall 14	34	37	71
	Spring 15	17	21	38
Total		51	95	146

Table 6.3 Recursion Experiments Test Scores and Learning Gains

School	Semester	Condition	Pre-test		Post-test		Gain	
			μ	σ	μ	σ	μ	σ
CMUQ	Fall 14	System	5.43	1.71	6.03	1.33	0.6	1.43
	Fall 15	System	5.14	2.01	6.43	2.09	1.29	1.75
UIC	Fall 14	Paper	4.75	2.39	7.25	1.37	2.5	0.54
	Fall 14	System	6.74	2.59	7.46	1.90	0.72	2.20
UIC	Spring 15	Paper	7.53	2.35	8.94	1.30	1.41	1.97
	Spring 15	System	7.19	3.01	7.05	2.16	-0.14	2.89

6.2.3.2 Experiments at CMU Qatar

The students from CMUQ were taking the 15-112 class, *Fundamentals of Programming and Computer Science*. This introductory course focuses on Python programming, and is taken by non-majors, such as Information Systems, Biology, and Business. In this class, recursion was covered over two consecutive weeks; the students participated in the experiment during the second week, namely, while they were still being taught recursion.

If all CMUQ students are considered together (N=37), the difference between pre-test ($\mu = 0.527, \sigma = 0.18$) and post-test ($\mu = 0.626, \sigma = 0.18$) is significant ($t = 3.9, df = 36, p = 0.001$) - please see Table 6.3. Probing deeper, this is due to a significant difference between pre- and post-test for the Fall 15 experiments ($t = 3.28, df = 20, p = 0.004$); no such difference was found for Fall 14, perhaps due to a smaller number of students that semester.

6.2.3.3 Experiments at UIC

The students at UIC were all enrolled in the CS151 class, *Mathematical Foundations of Computing*, where recursion and induction are introduced. This course is required for CS and Computer Engineering majors, and is restricted to students in the College of Engineering. In both the semesters, induction first and recursion second were covered in 4 consecutive weeks, from about week 4 to week 7 in a 15 week semester. However, because of logistics reasons, the experiments took place at different points in the two semesters: in Fall 2014, about two weeks after the end of the lectures on recursion, but in Spring 2015, about seven weeks after that same point in the class coverage (namely, 5 weeks later than the experiments in Fall 2014).

The UIC students were assigned to condition by laboratory session: in Fall 2014, of 4 laboratory sessions during the day, the first one and last one were assigned to the paper condition and the middle two to the system condition; in Fall 2015, of the 5 laboratory sessions, the first one and the last two were assigned to the paper condition and the middle ones to the system condition. The main goal of these assignments was to partition the subjects in two groups of similar size.

In the Fall 2014 experiments at UIC, we observe a highly significant difference between pre- and post-test, if the students are taken all together - please see Table 6.3. The difference between pre-test ($\mu = 0.58$, $\sigma = 0.27$) and post-test ($\mu = 0.74$, $\sigma = 0.16$) is highly significant ($t = 4.1982$, $df = 70$, $p < 0.0001$). On deeper analysis, however, it is the paper condition where learning occurs: the difference between pre- and post-test is significant for the paper condition ($t = 7.5050$, $df = 33$, $p < 0.0001$), whereas it is not significant for the system condition. In fact, there is a significant difference between the pre-tests of the paper and system conditions in Fall 2014 ($t = 3.3418$, $df = 69$, $p = 0.0013$). In these circumstances, observing the difference between normalized gains between the paper and system condition is more appropriate, and this is significant as well ($t = 2.7611$, $df = 70$, $p = 0.0074$).

In the Spring 2015 experiments at UIC, there are no differences in pre-test between paper and system, but again, the paper condition does better as concerns both absolute gain (a trend, $p = 0.07$) and normalized gain ($p < 0.05$).

If we consolidate the UIC experiments, we still see a significant difference between pre- and post-test for the paper condition ($t = 7.6308$, $df = 50$, $p < 0.0001$); additionally, there is a significant difference between the gains in the paper condition and the system condition both for absolute gain ($t = 3.9861$, $df = 107$, $p = 0.0001$,) and for normalized gain ($t = 3.3396$, $df = 107$, $p = 0.0012$).

The conclusion from these experiments is that, at UIC, the paper condition is apparently doing better than the system condition, even taking into account the very low pre-test score for the paper condition in Fall 2014. Since the students came from the same course, it appears the difference in pre-tests among conditions is an effect of pure serendipity. However, the paper condition is basically equivalent to the system condition in terms of types of activities the students go through, other than watching the animation of the algorithm execution. In fact, a better analysis of these experiments is in terms of the correlation of the activities the students perform and their learning gains, which is what we will discuss in the next section. Unfortunately, the activities done on paper during the paper condition have been lost, probably due to

Table 6.4 Recursion Experiments, System Only Condition - Revised
Student Distribution

Semester	School	# Students
CMUQ	Fall 14	16
	Fall 15	19
UIC	Fall 14	33
	Spring 15	21
Total		89

multiple moves of offices and laboratories, and we are unable to carry out any additional analysis for the paper condition.

6.2.4 Analysis of Students' Interactions with the System

We will now focus only on the system condition at the two campuses, to analyze features of the interaction of the students with the system, in order to uncover which tasks (if any) are conducive to learning. For this analysis, we had to exclude few subjects (4 from the UIC Fall 14 group, and 2 from the CMU Fall 15 group), because no system logs exist for them – they took the pre- and post-tests, and did use the system, but either they did not consent to their data being logged, or a logging malfunction resulted in no log being available. As a consequence, our analysis of the system condition will include 35 students at CMU and 54 at UIC, for a total of 89 (as opposed to 95 in Table 6.2). Table 6.4 provides the number of subjects to whom the analysis we present next applies.

The two groups of students at CMUQ and UIC significantly differ as concerns their initial knowledge as conveyed by the pre-test (CMUQ: $\mu = 0.52$, $\sigma = 0.19$; UIC: $\mu = 0.68$, $\sigma = 0.28$; unpaired t-test, $t = 7.5050, df = 87, p < 0.0023$). However, no significant differences exist between the two groups regarding absolute or normalized learning gains.

As we did in our analysis of the human–human tutoring dialogues in Chapter 3, we examined the correlations between the activities the students perform in ChiQat-Tutor and learning, to inform future research on which activities should be emphasized in the system. As in our earlier analyses, we added new variables to well-known variables that often have an effect on learning such as time on task, and pre-test. In fact, we started by analyzing the effect of time on learning, where time on task is the total time students interacted with the system.[1] The reason we explored time as a predictive variable is the existence

---

[1] This time is the difference elapsed between a student logging in and a student logging out, possibly summed over multiple login episodes. We do not have more specific measures available, to ascertain how much of this total time is effective time in which students were truly using the system. Whereas the various active tasks students engage in are logged in the system, just summing up their times would be inaccurate, since students are still productively interacting with the system when they read a problem, read feedback, or

of a significant difference in time spent on the system between students at CMUQ and UIC (CMUQ: $\mu = 25'30''$, $\sigma = 4'06''$; UIC: $\mu = 14'26''$ $\sigma = 6'01''$; unpaired t-test, $t = 9.1412$, $df = 87$, $p < 0.0001$). Whereas in principle all students had 30' to work on the system, at CMUQ experiments took place during 75' long formal lectures, at UIC during 50' long laboratory sessions. Presumably, students take more care being on time for a formal lecture than a laboratory session. More importantly, the lectures at CMU were 35' longer than the 50' nominal duration of the experiment, while the 50' lab sessions at UIC lasted the exact length of the whole experiment (10' pre-test + 30' system usage + 10' post-test). Clearly, there was wiggle room at CMUQ to accommodate late students and make sure that they all started using ChiQat-Tutor at about the same time (as the logs confirm). No such wiggle room was possible at UIC and in fact notes taken by the experimenter at the time, confirm that some students were delayed since they needed help to register with the system and/or start ChiQat-Tutor. At UIC, for many students, the interaction with ChiQat-Tutor started at around 20' after the hour, if not later (all these lab sessions started on the hour); this would have left them only 20' total, if not less, to use the system, since they were interrupted at 10' from the end of the lab session to have enough time to take the post-test.

We found that time correlated neither with learning gains, nor with normalized learning gains (a linear regression analysis between time spent on ChiQat-Tutor and learning gain was not significant; neither was a linear regression analysis between time on the system and normalized learning gain); only one linear regression between time and normalized learning gains, the one for CMUQ students, showed a trend toward significance. After ascertaining that time per se was not explanatory, we ran systematic multiple linear regressions which included time, pre-test, and the different tasks and subtasks students are engaged in. The original tasks were presented in Table 6.1; some noteworthy subtasks of the original tasks are listed in Table 6.5.[2]

The results of our regression analyses are reported in Tables 6.6 and 6.7. Analogously to how we reported results of multiple regressions in Chapter 3, we include: model A which only includes time on task and pre-test (as we just discussed, time per se is not a significant variable, other than the trend for the CMUQ students, for normalized gain); and then the best significant model B that adds other variables to time and pre-test, that are significantly correlated with absolute, or normalized, learning gain. In general, Models B have higher adjusted R^2 than the respective models A (the exception is for absolute gain, when considering all students together, see Table 6.6). For absolute gains, Models A and Models B are not pairwise significantly different;

---

study the graphical representations. For this type of more detailed analysis, eye tracking equipment would be necessary, but we had none at our disposal.

[2]While the subtasks from Table 6.5 are subsets of the corresponding main tasks, and hence collinear with the main task in multiple regression parlance, the regressions we ran include both tasks and subtasks, since the collinearity coefficients were always well below the VIF=5 tolerance threshold (Variance Inflation Factor).

Table 6.5 Tasks & Subtasks

Task	Subtask	Description
Answer	Answer Correct	# correct answers
Play animations	No subtasks	
Trace	Trace Correct	# tracing tasks correctly completed
	Node Trace Correct	# nodes clicked in correct order
	Node Trace Errors	# nodes students clicked in incorrect order
Validate	Validate Correct	# validating tasks correctly completed
Construct	Construct correct	# constructing tasks correctly completed
	Construct validation attempts	# validation attempts while constructing RGraphs

for normalized learning gains, Models B are pairwise, significantly better than the corresponding Models A counterparts, for all groups of students (only CMUQ, only UIC, all together).

Based on the results in Tables 6.6 and 6.7, we can offer some informed speculations on which tasks or subtasks may be conducive to learning. We use the term *speculations*, rather than conclusions, because the β coefficients are small and because different variables affect learning in the two groups of students (CMUQ and UIC), namely, we cannot truly generalize across the two groups of students.

First, pre-test is always included, as a measure of previous knowledge. The reader may be surprised that, in the regression for normalized gain, pre-test is included even if pre-test is used to compute normalized gain to start with: however, we found that pre-test functions as a negative confounder whose absence suppresses the (weak) effect of other factors on learning [72].

The first generalization across our results is a negative one: no models include either the *construct* task, or any of its subtasks. Whereas there are few construct tasks in general (2 on average), it is not different from the average number of *play animations* (1 on average) or of *trace* (2 on average – *trace* intended here as a complete tracing task).

There are hardly any significant correlations for all students taken together. The only task or subtask that trends toward significance is the number of *play animations*, which is significant only for absolute gain; surprisingly, the correlation is negative. This negative (albeit tentative) result as concerns animation fits with the general assessment we discussed in Section 6.2.1.1: results were contradictory and indicated that simply watching an animation of an algorithm does not positively affect student learning.

For CMUQ students, no significant correlations are found in Model B for absolute gains; on the other hand, the number of *Play animations* negatively

Table 6.6 Regression Models: Absolute Gain as Dependent Variable

Students	Model	Predictor	β	R^2	P
CMUQ only	A	Time	0.052	0.212	ns
		Pre-test	−0.439		0.004
	B	Time	0.055		ns
		Pre-test	−0.411	0.223	0.004
		Answer	−0.058		ns
UIC only	A	Time	0.043	0.496	ns
		Pre-test	−0.653		< 0.001
	B	Time	0.049		ns
		Pre-test	−0.675		< 0.001
		Answer	0.154		0.014
		Play Animations	−0.218	0.564	ns
		Trace	−0.083		ns
		Node Trace Errors	0.063		ns
		Validate	−0.160		0.086
		Validate Correct	0.312		ns
All	A	Time	0.003	0.421	ns
		Pre-test	−0.573		< 0.001
	B	Time	0.019		ns
		Pre-test	−0.599		< 0.001
		Answer	0.033		ns
		Play Animations	−0.219		0.051
		Trace	−0.233	0.421	ns
		Node Trace Correct	0.020		ns
		Node Trace Errors	0.056		ns
		Validate	−0.096		ns
		Validate Correct	−0.096		ns

correlates, and the number of nodes correctly clicked during a tracing task positively correlates with normalized gains.

For UIC students, Model B for absolute gains has the highest explanatory power, with an adjusted $R^2 = 0.564$. In this model, the number of *answer* tasks positively correlates with learning gains; the number of *validate* tasks is negatively correlated with gains, but it is only marginally significant. As concerns normalized gains, Model B has much lower explanatory power, but still, the number of *answer* tasks positively correlates with gains, and so does the number of correctly completed *validate* tasks.

If we take these results as pointing to a way forward for further experiments (given they are weak for the reasons mentioned above), we would most likely eliminate the *construct* task, that in general did not appear to correlate with learning; and eliminate or at least modify the *play animation* task, since in the current version of the system, it is a passive task. Additionally, tracing an RGraph provides some of the same information that animations highlight,

Table 6.7 Regression Models: Normalized Gain as Dependent Variable

Students	Model	Predictor	β	R^2	P
CMUQ only	A	Time	0.023	0.191	0.069
		Pre-test	−0.075		0.023
	B	Time	0.009	0.269	ns
		Pre-test	−0.093		0.008
		Answer	−0.007		ns
		Play Animations	−0.061		0.046
		Trace	−0.026		ns
		Node Trace Correct	0.014		0.033
UIC only	A	Time	0.007	0.103	ns
		Pre-test	−0.118		0.023
	B	Time	−0.014	0.212	ns
		Pre-test	−0.145		0.002
		Answer	0.073		0.012
		Play Animations	−0.032		ns
		Trace	−0.015		ns
		Node Trace Errors	0.027		ns
		Validate	−0.058		ns
		Validate Correct	0.237		0.012
All	A	Time	0.007	0.137	ns
		Pre-test	−0.105		< 0.001
	B	Time	0.010	0.155	ns
		Pre-test	−0.111		< 0.001
		Answer	0.022		ns
		Play Animations	−0.043		ns

but students are more active since they need to click on nodes in the correct order of execution, instead of simply watching the animation play.

6.3 SUMMARY

In this chapter, we have presented additional lessons that we have developed for ChiQat-Tutor for other important introductory topics in Computer Science: Binary Search Trees (BSTs) and recursion. We ran a pilot, in-laboratory evaluation for BSTs; and we ran a more extensive and ecologically valid evaluation for recursion, in line with the evaluation for the various versions of ChiQat-Tutor we developed for linked lists. Whereas the evaluation for the recursion module was not as explanatory as the evaluations we ran for the linked list module, it provides us with a path forward for further experimentation.

While the lesson for BSTs was a direct extension to our previous work on linked lists, recursion required a reconceptualization of the teaching strategies themselves: the focus was on understanding recursion per se, as a process, as opposed to manipulating data structure, where the focus was more on the state of the data structure. It is important to note that the architecture of ChiQat-Tutor, described in Chapter 4, supported the development of the recursion module as is, which attests to its flexibility. The next chapter will illustrate in complete detail how such architecture can be used to develop new lessons from scratch.

CHAPTER 7

A Practical Guide to Extending ChiQat-Tutor

CONTENTS

U P to now, we have discussed the work we have done with ChiQat-Tutor, from a variety of perspectives. It is now time to look to the future, and specifically, provide our readers with insight into extending ChiQat-Tutor. ChiQat-Tutor is a live system that we hope many readers will download and take advantage of. It is available at http://www.digitaltutor.net/.

Extending ChiQat-Tutor can be done along several dimensions, such as adding new domain content, or adding new pedagogical strategies. All of these elements are implemented in ChiQat-Tutor as *plugins*. In this chapter, we will present a full example for new plugin development; specifically, we will show how to define a new lesson on stacks. By the end of this chapter, the reader will have sufficient knowledge, and practical guidelines, to integrate new components into the ChiQat-Tutor framework. On the other hand, readers who are not interested in implementation details can safely skip this chapter and jump to the conclusions in the next chapter.

7.1   AN IMPLEMENTATION ARCHITECTURE

In Chapter 4, we presented the architecture of ChiQat-Tutor at an abstract level (see Figure 4.1). We will now dive into how that abstract architecture is implemented in practice.

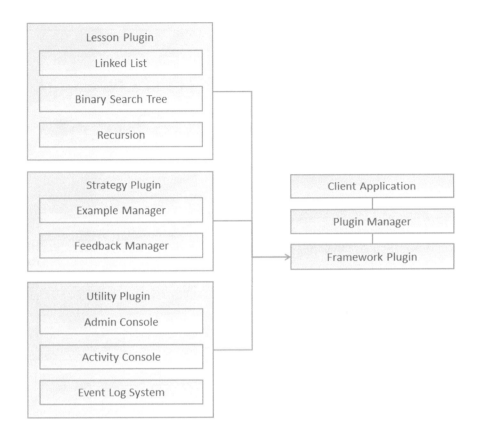

Figure 7.1 Implementation: Plugin architecture.

To start with, we will note that ChiQat-Tutor is a modular desktop application, with sparse communication between the clients and a central server. To keep as much processing as possible local to the user, the majority of the architecture is concentrated on the client side application. The server side is minimized to act as system coordinator and data store.

The core of the system is in the client side application, whose strength is in its modular architecture: components can be created independently, and can communicate together for a high degree of software reuse. The implementation structure of the client side application is depicted in Figure 7.1.

There are two main components in the client application: a host application and plugins. The host application acts as the entry point and glue of all client components; application specific functionality is encapsulated in plugins.

All plugins are extended from a common plugin type. There are three general types of plugins: lesson, teaching strategy and utility plugins. Each plugin can communicate with other plugins via an internal messaging API. On execution, the host application dynamically finds all applicable plugins, and automatically loads and initializes them. This process of plugin selection is invisible to the end user, yet it is configurable by the system administrator.

If we compare the implementation level architecture in Figure 7.1 to the higher level architecture we presented in Figure 4.1, it is apparent that the Tutor Module from Figure 4.1 maps quite directly to the *Strategy Plugin* in Figure 7.1, with the two submodules "Feedback manager" and "Example manager" directly mirrored as subplugins. The other higher level components from Figure 4.1 are not directly mapped in Figure 7.1 but are spread across various plugins. For example, Problem and Solution Definitions from the Domain Model will be included in one of the corresponding lesson plugins (Linked List, which includes the PKM as well; Binary Search Tree; or Recursion). Finally, the User Interface is rendered by code implemented in the corresponding lessons and in auxiliary Utility plugins as appropriate.

7.2 CASE STUDY: THE STACK TUTOR PLUGIN

In this section, we will illustrate how to design and implement a new lesson plugin for ChiQat-Tutor. We will do so by working on a case study: a brand new "Stack Tutor" plugin. This plugin, although relatively simple, is fully functional, and the reader can use it as a starting point to develop their own plugins. In this section, we assume that the reader has a good working knowledge of the Java programming language; experience with the JavaFX framework for developing graphical user interfaces and functional knowledge of the Stack data structure (see Appendix A for a refresher), which is the domain subject of the plugin we are going to present here.

In this implementation, we chose to utterly simplify the pedagogical strategies we use with respect to what we discussed in the rest of the book, in order to limit the complexity and length of the presentation in this chapter. Additionally, we illustrate choices such as providing a correct solution to the student on demand, which was not available in the versions of ChiQat-Tutor-v1 and ChiQat-Tutor-v2 we experimented with. This was done to demonstrate the flexibility of ChiQat-Tutor's modular architecture, which is agnostic as concerns the pedagogy behind the specific lesson, and allows the instructor / developer to fit it to their purpose.

7.2.1 Stack Plugin Design

The purpose of the Stack Plugin is to allow students to practice the canonical operations on a stack data structure. The three main operations on a stack are the following:

1. *push(X)* adds a new element X to the top of the stack.

2. *pop()* removes the element currently at the top of stack, and returns it.

3. *top()* returns the element currently at the top of the stack, but does not remove it.

A type of exercise that students can do to practice with stacks is the following: given a sequence of operations involving a stack and an optional set of other variables, trace the content of the stack and each variable after the execution of each operation. This type of exercise and its solution can be represented in a tabular format. For example, in the pre-/post-test for our human tutoring experiments, included in Appendix B, we asked the students to solve the following problem (repeated here for convenience):

You have a stack data structure. The operations defined on it are push, pop, and top, with the usual semantics for stacks.

Your stack is initially empty. Write down the state of the stack and the content of the variable x after executing each of the following operations.

Operation	Stack	x
x = "A";		
push("B");		
push("C");		
push(x);		
pop();		
x = top();		
push("A");		
push(top());		

Here is the solution for this problem:

Operation	Stack	x
x = "A";		A
push("B");	B	A
push("C");	B C	A
push(x);	B C A	A
pop();	B C	A
x = top();	B C	C
push("A");	B C A	C
push(top());	B C A C	C

Our Stack Plugin should present students with problems of this type. Students should be able to work on the problems using a tabular graphical interface, and the system should be able to evaluate the correctness of a student's solution. Moreover, the system must be capable to automatically generate an unlimited number of new problems with a level of complexity specified by the student. In the versions of ChiQat-Tutor we discussed earlier in the book, the student was not allowed to ask ChiQat-Tutor to generate new problems on the fly. Beyond demonstrating ChiQat-Tutor's architecture flexibility, this choice makes sense for stacks. The rationale is that for stacks, problems solved by one single pop or push operation on the stack are too simple. Learning stacks requires to become proficient with any combination of these operations; hence, our hypothesis is that being able to vary the parameters of the problem and hence solving more varied problems should be conducive to better learning. More in general, automatic generation of problems relieves instructors of the burden of authoring those problems, and students can keep practicing as much as they want without the system running out of problems [344].

Upon starting the Stack Plugin, its interface will include the following components – note that the user below can be the students themselves, as just discussed, or the instructor:

- An input field where the user will enter the desired number of operations (steps) of the problem to be solved. In our previous example the problem had 8 steps.

- An input field where the user will enter the number of variables (in addition to the stack) that the problem will have. In our previous example, the problem had 1 variable (named x).

- A button that will submit a request to the system to generate a new problem.

After the system generates a new problem, the interface will add the following components:

- A table with a header, and a row for each step of the problem. Each row should indicate the step number, the operation to be executed on that step and a sequence of input fields where the user will write the content of the stack and the value of each variable included in the problem.

- A button that will submit the student's solution to the system. Upon submission, the system will check the solution for correctness and provide color-coded feedback: the correct parts of the student's solution will turn green; the incorrect parts will turn red. Very simple verbal feedback will be provided as well, to summarize the answers' correctness, as in *Correct: 5/15 (33.3%)*.

- A button that will reveal the correct solution to the student, which, as noted earlier, was not available in the lessons on linked lists, BSTs and recursion. Still, for pedagogical reasons, we want to disable this button until at least one solution attempt has been submitted by the student.

Figure 7.2 shows an example of the interface we just described.

7.2.2 Class Structure

Our Stack plugin is composed of five Java classes and one CSS (Cascading Style Sheet) file; all the code necessary to implement the plugin can be found in Appendix E.

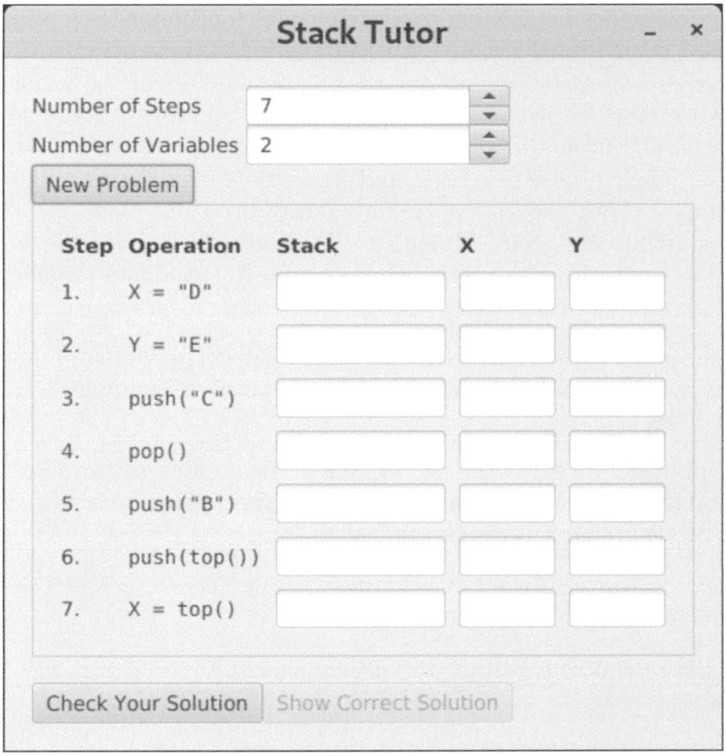

Figure 7.2 Main Interface of the Stack Plugin.

PluginInstance.java This class needs to be implemented in order to connect the plugin to the rest of the ChiQat-Tutor architecture, but it does not contain any of the logic of the actual plugin.

StackView.java This class implements the graphical interface of the Stack plugin, as well as some of the behaviors of the plugin upon activation of the corresponding buttons.

StackProblem.java This class represents a stack tracing problem. It contains the logic to automatically generate a new randomized problem; to simulate the canonical stack operations (push, pop, top) and to check the correctness of a given solution to the problem.

StackProblemStep.java This class represents a single step for a stack tracing problem.

StackProblemFeedback.java This class represents the result of the comparison between a submitted solution and the correct solution for a stack tracing problem.

`StackStyle.css` This Cascading Style Sheet (CSS) includes configuration parameters for the graphical interface of the plugin, such as fonts and spacing between components.

7.2.3 Setting up the Stage

The first class that we need to implement is named `PluginInstance`. The purpose of this class is to connect our new plugin to the rest of the ChiQat-Tutor framework. We can simply copy the bulk of this class from our template, and update it at lines 35–36 with the name of the class representing the entry point of our plugin's user interface (the "view"), which we will discuss in the following section (please see Section 1 in Appendix E).

7.2.4 Graphical Interface

The implementation of the graphical user interface (GUI) of our plugin is named `StackView`, which extends the class `PluginViewFx` from the ChiQat-Tutor framework. It contains a variety of JafaFx graphical input/output elements (text fields, spinners, labels, buttons), and crucially a reference to an instance of the `StackProblem` class (discussed in the next section) which implements the main logic of the plugin.

The method `Init()` is the first method executed by the ChiQat-Tutor framework upon activation of the plugin. In our case, the main initialization involves the creation of a root pane, which is delegated to the method `createMainPane()`. This method instantiates the graphical widgets, and sets their callback methods where appropriate (such as `createProblem()`, `checkSolution()`, and `showSolution()` for the three buttons). After the root pane is created, it is connected to a CSS style sheet which sets several cosmetic parameters, such as font style and size, and spacing between components.

The method `createProblem()` is called when the student presses the "New Problem" button. The method reads the current state of the widgets containing the configuration parameters for the new problem to be created, such as the number of steps (i.e., the length of the problem), and the list of variable names to be added to the problem. It then invokes the constructor for `StackProblem` (described in the next section). After the new `StackProblem` object is created by the constuctor, this method completes the graphical interface by adding a collection of input/output widgets representing the problem. Our stack problems are well represented in a tabular format; the graphical widgets are laid out as such (Figure 7.2).

The method `checkSolution()` is invoked when the student clicks the "Check Your Solution" button. Checking a solution involves three steps: (1) collecting the current input from the student, which is spread across the various text fields of the table; (2) passing the student input to the `checkSolution()` method of class `StackProblem` and (3) visualize the result of that method in the form of color-coded feedback (correct parts of the solution turn green, whereas wrong elements turn red), and include a verbal feedback message as well.

The method `showSolution()` is called upon by clicking the "Show Correct Solution" button. It displays the correct solution of the current problem to the student upon request. The "Show Correct Solution" button is disabled until the student tries to submit his/her own solution attempt at least once.

7.2.5 Stack Problem Logic and Feedback

The logic of the plugin is coded in three classes: `StackProblem`, `StackProblemStep` and `StackProblemFeedback`.

`StackProblemStep` is a passive data structure (it does not include any method) which contains the three components of each step of a stack tracing problem:

1. An operation (either push, pop, top, or assignment) to be applied to the previous state of the stack and/or any of the variables in the problem.

2. The content of the stack after the execution of the operation.

3. The content of the variables after the execution of the operation.

`StackProblemFeedback` is a class whose purpose is to generate simple textual feedback based on the correctness of a student's solution; hence, it implements *final feedback* (see Section 4.3.1), and only final feedback, of the types of feedback we discussed in Section 4.4. Its constructor takes as input a matrix of Boolean values which correspond to the correctness of each cell of the input table entered by the student. The constructor simply counts the number of correct items, and stores this number for later use. The actual textual feedback is generated by the method `getFeedback()`: this method composes a simple one-line report of the number of correct items, expressed both as an absolute number and as a percentage of correct items over the total possible items. Notice that this text is not the only form of feedback students receive: students also get the color-coded feedback described in the previous section, handled directly by the graphical user interface.

The `StackProblem` class implements the heart of the logic of the plugin. Its constructor method is responsible for creating a new randomized problem based on the size parameters specified by the user: number of steps, and number of auxiliary variables (see Figure 7.2 for an example of a randomly generated problem with 7 steps and 2 variables). The `StackProblem()` constructor also allows to customize the names of the variables (X and Y in the figure) and the problem alphabet (possible values that can be stored in the stack and/or variables, such as "B", "C", "D" and "E" in the figure); however, we chose not to include these as configurable parameters in the user interface, so the possible variable names and alphabet are fixed in the constant fields `VAR_NAMES` and `ALPHABET` of class `StackView`, respectively.

The random problem-generation process starts by adding a sufficient number of assignment operations to the problem steps so that all the variables in the problem will be initialized with some value. Then, the following steps to be added are chosen randomly among stack operations (push, pop, top) or variable assignments. The random selection process is controlled in such a way that no sequence of operations could possibly lead to an error state in the problem. As soon as each step is generated, the correct outcome of that step is stored in the corresponding `StackProblemStep` object. This will allow for a quick comparison between the correct solution and the student's solution.

The three methods `push()`, `pop()` and `top()` simulate the effect of the corresponding stack operations on the current problem's stack state (which is internally represented as a simple string).

Finally, the method `checkSolution()` compares a given student solution (i.e., the full content of the input table shown in the user interface) to the correct solution, which was pre-computed at the time of problem generation. The result of `checkSolution()` is a matrix of Boolean values that identify which cells of the table have been filled out correctly by the student. This result matrix will be used by the user interface to turn all the correct cells to green, and the incorrect ones to red; and by the feedback generator (a `StackProblemFeedback` object) to compose an appropriate text message for the student.

Conclusions

CONTENTS

W E have come to the end of our journey into providing supportive educational technology for introductory Computer Science at the college level. We hope the reader will have found our research inspiring and useful, and will be able to build on it. As we mentioned earlier in the book, ChiQat-Tutor is available at http://www.digitaltutor.net/. The annotated tutoring data is also available at http://www.digitaltutor.net/chiqat/pages/database.php.

In this chapter, we briefly summarize our approach, and the contributions of our work; and discuss some of the many possible venues for future work.

8.1   WHERE WE ARE, AND LESSONS LEARNED

The research journey discussed in this book lasted about 15 years. It is grounded in a keen interest in cognitive theories of learning, and specifically in how to explain what happens, cognitively and linguistically, in one-on-one tutoring; in how to computationally model the findings from our tutoring dialogues in ChiQat-Tutor, our ITS; and in experimenting and validating those findings in the actual deployment of a number of versions of ChiQat-Tutor.

We started with two prongs into the problem: the first from cognitive science, specifically, Ohlsson's theory of learning (Ohlsson is in fact a contributor to several chapters in this book), and the second from Computational Linguistics, how to annotate and analyze human-human conversations. Naturally, our third prong comes from the profession that two of us (Di Eugenio and Fossati) have chosen, that of Computer Science educators: hence, our tutoring dialogues were collected in the area of introductory data structures (linked lists, stacks and BSTs).

In Chapter 3, we extensively described our data collection, and how we transcribed and annotated the data, with coding categories that reflected both the tenets of the theory of learning we espoused, and pragmatic notions of conversation. We carefully validated our annotations with multiple rounds of intercoder agreement measurements. Once the data was annotated for meaningful categories, we explored how those categories correlated with learning. We discussed many regression models, which always include pre-test as a co-variate: we started with utterance level categories, i.e., moves, then with sequences of these categories, and finally, with the episodic strategies of worked out examples (WOEs) and analogies. We found that the most explanatory models resided either with sequences of individual moves – for example, for lists, the sequence *[PT, DPI, FB]* (a sequence of three moves, *Prompt* followed by *Direct Procedural Instruction* followed by *Feedback*); or with features of the episodic strategies that made use of the utterance level categories – for example, for stacks, a model where the number of FB+ within a WOE (*WOE-FB+*) positively correlates with learning gains, while the number of prompts within WOEs (*WOE-PT*) negatively correlates with learning gains (please see Section 3.5 for the summary).

Finding pedagogical strategies that correlate with learning is necessary definitely not sufficient for computational models to be embedded in an ITS. From a theoretical point of view, our regressions only show which strategies correlate with learning, but not under which conditions those strategies are applicable; as we noted in Chapters 3 and 4, no human tutor would mechanically go through say, cycles of *[PT, DPI, FB]* even if as we just pointed out, this is an effective pedagogical strategy for lists. From a practical point of view, an ITS is a complex software artifact that needs to be developed according to sound software engineering principles.

We provided insight into these issues in Chapter 4, in which we described the reusable plugin architecture we adopted for ChiQat-Tutor and its main three components (Domain Model, Tutor Module and User Interface). We also illustrated the processing that ChiQat-Tutor performs when interacting with a student, as a way to highlight the conditions under which we provide different types of feedback. In turn, *reactive* and *proactive* feedback are in part modulated according to the *Procedural Knowledge Model (PKM)*, a probabilistic graph that for each problem, succinctly represents all the correct and incorrect paths that students took while attempting to solve that specific problem. A first version of the PKM was automatically trained on the logs of the students who used ChiQat-Tutor-v1.1 and ChiQat-Tutor-v1.2, to be used by the ChiQat-Tutor-v1.3 through 5 versions; at that point, the PKM was retrained using the logs of all students who had used the five versions of ChiQat-Tutor-v1, and this final PKM was used by ChiQat-Tutor-v2. During interaction with a student s, ChiQat-Tutor continuously matches s's actions and the current state of problem solving to the PKM; depending on the features of the matched state, and other features of the interaction such as uncertainty level, and time from last action, ChiQat-Tutor provides reactive or proactive feedback (or no feedback).

Chapter 5 was devoted to several rounds of evaluation, first of the five versions of ChiQat-Tutor-v1, which differed as concerns the feedback they provided, and then of the six versions of ChiQat-Tutor-v2, which differed according to which type of WOE and/or of analogy they provided. ChiQat-Tutor-v1 was evaluated with 219 students, and ChiQat-Tutor-v2, with 313.

ChiQat-Tutor-v1 was evaluated only in introductory classes required of majors, at UIC and USNA (the US Naval Academy). There was a progressive improvement in learning gains from ChiQat-Tutor-v1.1 to ChiQat-Tutor-v1.5, with the learning gain in ChiQat-Tutor-v1.5 being identical to the learning gain with human tutors; however, there were no significant differences among the five versions. The only significant differences with respect to a control group were with the human tutored group, and with the ChiQat-Tutor-v1.5 group. Additionally, on the whole, the triad ChiQat-Tutor-v1.3, v1.4 and v1.5, which provide the more complex reactive and proactive feedback, engender better performance from the students as concerns number of problems solved, and goodness of their paths to solutions.

ChiQat-Tutor-v2 was evaluated with both majors and non-majors. First, there are significant differences concerning learning gains between majors and non majors; the learning gain for non majors, across all conditions ($\mu = 0.14$) is identical to the learning gain with ChiQat-Tutor-v1.5, which, in turn, is identical to the learning gain with human tutors. Somewhat unexpectedly, we did not find significant differences among conditions, neither with majors nor with non-majors. We did observe that learning gains in time constrained conditions are the lowest, even if they are not significantly different from learning gains in other conditions (there were two such conditions, *NoExit* in which the student must step through the WOE till completion, and *TimeOut*, where a time limit is imposed on each WOE, so that the student will not be able to complete it). Additionally, given the observation regarding non-majors learning more, we divided the students into beginners or advanced, independently of major, and we found a trend toward significance: advanced students learn more from short WOEs than from long WOEs.

The last finding we just mentioned spurred us to investigate how to automatically establish whether a student is beginner or advanced, given their behavior on the first problem in ChiQat-Tutor-v2: the rationale is that all students would be served a standard WOE in that first problem (if they requested it) and, for the remainder of the problems, would be provided a short or standard WOE, according to ChiQat-Tutor-v2's assessment of their knowledge level. We used machine-learning algorithms on the logs, and found our best models to use features related to the number of correct submissions, the number of positive feedback messages (which is consistent with the results on the effectiveness of positive feedback in ChiQat-Tutor-v1.3 through ChiQat-Tutor-v1.5), and the time spent on certain specific steps in the WOE for Problem 1, for example, Step 7 that actually suggests a wrong move, and Step 9 that explains why step 7 is wrong.

Chapters 6 and 7 are devoted to extensions to the approach embodied in ChiQat-Tutor. Chapter 6 starts with the BST module. This extension is organic, in that it only required definitions of new lessons on BSTs and the relevant problems, and minor updates to the interface to manipulate a two-dimensional structure, as opposed to the uni-dimensional lists. More challenging was the extension to recursion, in that the focus was on understanding recursion per se, as a process, as opposed to manipulating a data structure, where the focus is more on the status of the data structure. We devised and implemented new pedagogical strategies specific to teaching recursion, such as manipulation and processing of Recursive Graphs (RGraphs). The flexible ChiQat-Tutor's plugin architecture was extremely effective in supporting the implementation of the recursion module.

The BST module was only piloted but not systematically evaluated; instead, ChiQat-Tutor-recursion was evaluated at two institutions, UIC and CMUQ (Carnegie Mellon University, Qatar campus). We found that students at CMUQ did learn when using the system. At UIC, but not at CMUQ, students were divided into two conditions (paper and system), and learned in the paper but not in the system condition. We then further analyzed correlations between learning and the tasks students were engaged with in the system. We did not find strong results, but some trends suggest that *Playing animations* negatively correlated with learning (perhaps because students are passive when watching an animation), whereas *validating* the RGraph had a small positive correlation with learning gains.

Finally, Chapter 7 is a practical guide to extending ChiQat-Tutor to new materials, using stacks as an example. Chapter 7 describes the full implementation at a higher level, with Appendix E providing the full code.

Besides the specific results we obtained, this decade-plus long research reinforced our belief that computational models of learning both need to be steeped in theoretical models of cognition, and must at least attempt to model the richness and complexity of human-human interaction. We also were further convinced that given the still vast gulf between human-human interaction and human-computer interaction, there is no guarantee that an effective tutoring strategy between a human tutor and a student is equally effective in a computer tutor: computational models are sometimes a very crude approximation to all the nuances that a human tutor uses in his/her craft, including a deep understanding of the student's mental model and emotional state.

Additionally, ecologically valid experimentation in the classroom brings with it factors that are beyond the experimenters' control and that may hinder generalization of results, if in fact they do not prevent experimenters from getting results at all. For example, our experiments with ChiQat-Tutor-v1.3 were affected by a severe network crash that forced the students to repeat the experiment about a week later, which is probably responsible for the very low level of satisfaction expressed by the students using ChiQat-Tutor-v1.3.

In the first session of evaluation of ChiQat-Tutor-v2, we had not yet linked the participation in the laboratory to a few points in the corresponding class, which may explain why in this particular session, the students learned nothing, as opposed to students always learning at least a bit in all the other sessions we conducted across ChiQat-Tutor-v1 and ChiQat-Tutor-v2. In the evaluation of recursion at UIC and CMUQ, the difference in session type and length (a 50' laboratory session at UIC, a 75' lecture session at CMUQ) may have affected the difference in results, in addition to the very different characteristics of the students in the two universities.

One answer to the serendipity of human subject experimentation is to run large studies as in medical research, where such differences may be washed out. However, this is not possible in an academic context, where resources are limited. The other option is to make the system available for free use, as indeed ChiQat-Tutor is, and to garner insight into its effectiveness by mining the logs of its users, a venue for future work to which we turn now.

8.2 FUTURE WORK

There are many potential venues for future work, some shorter term, some longer term.

8.2.1 Extending the Curriculum

The simplest extensions concern broadening the curriculum in ChiQat-Tutor, so as to cover all data structures presented in an introductory data structure class. The existing curriculum in ChiQat-Tutor already includes several such data structures, with the exceptions of graphs. Introducing graphs and problems on simple graph algorithms, such as breadth first search and Minimum Spanning tree, would be simple, at least in an iterative guise. As we showed in Chapters 6 and 7, it is easy to add new data structures and relevant problems to ChiQat-Tutor.

The next extension concerns a conceptual integration of the data structure modules with the recursion module. The attentive reader has probably noticed that the recursion module is different from the others, not just in appearance: neither do the algorithms it illustrates use data structures nor does the module itself rely on the pedagogical strategies we had discussed in the rest of the book. Naturally, the next step would be to more closely integrate the data structure curriculum with the recursion curriculum. Indeed, algorithms for many data structures can be written in a recursive fashion, and for trees and graphs, recursive, not iterative, algorithms directly correspond to their inherently recursive definitions.

Whereas the integration of these two foundational components of the introductory CS curriculum of ChiQat-Tutor-v2 is necessary, it is not trivial. Beyond careful planning of how concepts are introduced, since recursive problems on data structures cannot be provided before recursion itself is understood, the recursion module does not make use of feedback in the sense we discussed in Sec. 4.4, other than execution feedback for the final solution, and some color-coded feedback while playing animations or manipulating RGraphs. Neither does the recursion module utilize WOEs, although *playing an animation* could be considered as a graphical representation of a WOE. Indeed, not even the BST module provides reactive/proactive feedback or WOEs.

Reactive/proactive feedback is based on the PKM, which, in turn, is derived by mining the logs of students who have previously used ChiQat-Tutor. The size of the logs would be sufficient to develop at least a first version of the PKM for the recursion module, which was evaluated with 95 students; this does not hold for the BST module though, since it was only piloted with few students. We will come back to how small data can be augmented in Section 8.2.3. Additionally, proactive feedback requires alternatives for the student to choose between (see Figure 4.6 in Chapter 4); in the current implementation, one of these alternatives is correct, and the other three were manually chosen among the most common mistakes in the students' logs. We would need to explore how to automatically infer those alternatives from the PKM.

8.2.2 Enhancing Communication with the Student

One aspect that we never fully explored in ChiQat-Tutor is the student's input, which is currently limited. Students can input code in the *Student Code Input* window (the rightmost window in the ChiQat-Tutor's interface, see any figure illustrating ChiQat-Tutor, such as Figures 4.6, 4.7 and 4.8 in Chapter 4). Additionally, students can rearrange the nodes in the bottom left window (*State Space View*), and choose answers via radio buttons when provided with proactive feedback. Even if, as we discussed in Section 2.2, two-way interaction as in human conversation is one of the most effective ways of learning, we chose not to invest in Natural Language Understanding (NLU) for ChiQat-Tutor: true NLU would have required a massive effort on its own, and our limited resources did not allow us to do so. Additionally, the findings from our data collection indicated that more fully modeling the tutors first would be wise, since our tutors talk much more than students, producing more than ten times as many words as students do. Of course, as we noted in our analyses in Chapter 3, it is not what happens a lot that necessarily counts; for example, there are only 7.7 student initiatives (SIs) per session as opposed to 31.2 prompts per session on the part of the tutor, but SIs do appear in some of the more explanatory regression models, such as trigrams of dialogue acts for BSTs.

Adding true NLU to ChiQat-Tutor would then be the focus of one potential major next effort. As we discussed in Section 2.5, several ITSs have explored NLU to understand either students' answers, or to support students' activities such as self-explanation [4, 7, 90, 108, 111, 115, 134, 253, 254, 321, 367, 417, 428]. As we had noted in our discussion in Section 2.5, the sophistication in many of these approaches is more on modeling the tutoring conversation per se, than on understanding the specific input from the student – namely, the aim is to classify the student's input in a certain way (correct/incorrect/ uncertain, matching a known misconception or not, showing frustration), rather than fully understand what the student is saying. Even when the focus is on the unfolding of a tutoring dialogue, sometimes, the interaction is scripted, namely, the flow of the dialogue is hardcoded.

In some related work of ours on peer tutoring that we also described in Section 2.5, we approached NLU from similar points of view to what we just described. In KSC-PaL [202, 203, 226, 227], we approximated knowledge co-construction via *task initiative*. KSC-PaL used the same architecture of ChiQat-Tutor-v1 but was able to recognize whether the student has task initiative given their utterance or their action, using features including the dialogue act of the previous utterance, which was automatically labeled. The goal in KSC-PaL was to develop models in which the system itself could play the role of a peer, shifting back to a tutor role when the student really flounders.

In additional work, we explored how ChiQat-Tutor could support two human peers collaborating on the problems they solve with the system, rather than ChiQat-Tutor playing directly the role of a peer [182, 183, 185, 184]. We collected twenty 40' long conversations among dyads using ChiQat-Tutor-v2 (the version with standard WOEs). Two conditions were explored: in one, the only change to ChiQat-Tutor-v2 was the addition of a button that for each line of code, would attribute the line to one of the two participants; in the other, a graphical representation of the collaboration was added, which included a pie chart showing the number of contributions by each partner, where a contribution was a spoken utterance or a code submission. The conversations amount to over 9000 spoken utterances and over 40,000 words. We could use these conversations to drive models of the language students use when discussing linked lists, since each of these sessions includes approximately 2000 words, as opposed to 450 words per session when a student spoke with our human tutors.

Our focus on understanding the student input in this section does not mean that the other NLP side of interaction, generation, is solved in ChiQat-Tutor. Whereas all the pedagogical strategies we developed in ChiQat-Tutor use language, from various types of feedback to WOE to analogy, the specific wording of the feedback is hardcoded. Below, we will discuss how to make progress on automatically generating a more varied repertoire of expressions.

8.2.3 Mining the User Logs, and Deep Learning

We have mentioned the need to mine the ChiQat-Tutor's logs and/or additional conversational data, whether collected by us or by others, for further development, especially as concerns understanding students' inputs, and providing more varied expressions for feedback. This requires appropriate machine learning and data science approaches, which brings us to briefly touch on our potential adoption of the so-called "deep learning" methods such as the latest GPT3 [1] that have emerged in the last five years and have come to dominate Artificial Intelligence. Deep-learning models are based on massive neural networks, and are attractive since they have achieved state of the art performance on many tasks in NLP, even if their success does not necessarily translate into true understanding of the phenomenon they purportedly model [38]. Deep-learning models have started to be applied to educational data as well, especially when the data is abundant, for example log data, including comments, from student usage of MOOCs (Massive Open Online Courses) [9, 263, 437].

From our point of view, the first problem to address to use such models is what we call *the small data challenge*. Not just our data, but most available human tutoring data is tiny by deep learning standards. Many deep learning models are trained on huge amounts of data, such as the Common Crawl dataset which includes almost a trillion words [346]; in fact, one related problem is that they require enormous computational power, and hence, energy [395]. One venue of inquiry is to use the data we have collected as a seed for developing larger and possibly big data. The size of a corpus can be increased either by paraphrasing, which augments the original dataset with paraphrases of the original sentences [238, 427]; or by generating new documents by merging relevant segments within the documents with other irrelevant segments extracted from the corpus [62]. However, paraphrases, even when semantically equivalent, may subtly be different as concerns several features, such as affect; crowdsourcing may help in generating and/or evaluating paraphrases, but is not a panacea [211]. It is a research question whether the merging approach can be applied to conversations as opposed to documents. Another approach to increasing the size of data is to use active learning [380]: the algorithm interactively queries some information source to obtain the desired outputs for new data points that are automatically generated. In our case, to develop more expressions for feedback and for the steps in the WOE, and alternate between them, the information sources could be educators, learners, or additional annotated data, for example, other existing data in human-human tutoring.[2] Active learning has been used on its own in educational settings [187, 234] but not for tutoring dialogues.

Even assuming that we could effectively train deep learning models, many questions would need to be answered before we could reliably adopt them for a user-facing application such as an Intelligent Tutoring System. Deep-learning models are often black boxes with no explanatory power [363]: even when they perform better on say, a task such as assigning grades, they are not able to provide the reasons behind the specific score, which are needed to formulate feedback for the learner [431, 441]. Additionally, deep-learning models face serious issues of bias and unfairness [277], as has famously been

---

[1] https://www.nytimes.com/2020/11/24/science/artificial-intelligence-ai-gpt3.html, https://beta.openai.com/. Accessed Dec. 2020.

[2] For example, from repositories available at https://pslcdatashop.web.cmu.edu.

reported as concerns face recognition algorithms [60, 146]. In an educational domain, [23] shows that word embeddings trained on four subsets of college admissions essays, where subsets are determined by family income, perform differently on different standardized tasks, with the word embeddings trained on the essays of students belonging to the lowest income quartile, performing the worst.

Of course, all the issues we have just mentioned are research problems to be addressed, not obstacles that doom a research program. We are optimistic that the future of ChiQat-Tutor and of any other ITSs that engage students via language is bright, as long as the technology remains rooted in what we keep discovering on human cognition and communication.

A Primer on Data Structures

CONTENTS

A T a very abstract level, computers manipulate information, represented in a specific way. How data is best represented in computer memory depends on the processing that is to be done. In general however, almost any computer application needs to *store*, *retrieve* and *process* information. Information storage representations are usually called *data structures*. The procedures describing how the information is processed are known as *algorithms*. Among the most fundamental data structures are the three we focused on in our intervention: linked lists (lists for short), stacks and binary search trees (BSTs). Table A.1 summarizes features of these three data structures, including their dimensionality and the operations that they support, which we will describe shortly.

Table A.1 Data Structures and their Features

Name	Dimensions	Recursive	Algorithms		
			Search	Insert	Delete
Lists	One	Yes	Yes	Yes	Yes
Stacks	One	No	No	Yes	Yes
BSTs	Two	Yes	Yes	Yes	Yes

Among those three data structures, linked lists have strong ties to cognitive science. Lists are the foundation of the Information Programming Language (IPL) devised by Newell, Simon and colleagues [317, 318] and of Lisp [284], both used in Cognitive Science modelling and Artificial Intelligence. Lists are fundamental to the understanding of more complex representations in Computer Science, that are more appropriate for processing of different types. For example, a BST is the preferred representation or data structure to support a process for alphabetizing a list of words.

The order in which we introduce lists, stacks and BSTs is the order in which they are presented in most CS textbooks; it is also the same order in which the problems were presented to our subjects, both in our intervention and in the tests they took. These data structures are typically presented to students in one or two courses taken either during Spring of freshman year, or Fall of sophomore year.

A.1 LINKED LISTS (LISTS)

The main idea behind linked lists is that different pieces of information can be "linked" one after each other and then accessed sequentially. A common graphical representation of lists makes use of boxes and arrows (please refer to the drawing area in Figure A.1, repeated here from Figure 4.2). The unit of information in a list is called a *node*, and is represented by a box divided into two. A node contains the stored *data* (left half of the box) and a *pointer* to the following node (right half of the box). A *pointer* is an abstraction of the physical location of a node in computer memory. An arrow, starting from the pointer part of a box and ending at another box, represents the "link" between two nodes. To retrieve the information contained in a node, it is necessary to "follow the pointer."

The list L in Figure A.1 is composed of three nodes, containing the data *2*, *1* and *8*, respectively. Each node also contains the pointer to its successor, called `link` in this example (as shown in the multiple choice answers in the top middle feedback window). If a node has no successors, its pointer is set to null (indicated in Figure A.1 by the lack of an outgoing edge, as for the node containing *8* in its data part). Additionally, to access this sequence of three nodes in memory, we need to know where to start: this is what the variable L allows us to do. In the example in Figure A.1, a second variable called N points to a new node which will contain *3* and will need to be inserted between the first two elements of the list.

For any well-established data structure, a basic set of algorithms is commonly defined to effectively use the data structure itself. For linked lists, the most basic operations are *searching* for a certain element in the list – e.g., searching for *8* in the list in Figure A.1; *insertion* of a new node in the list – e.g., inserting *3* in the list in Figure A.1 by finding its right place after *2* and before *1*, as asked by the problem statement in the top left window, and *deletion* of an existing node from the list, for example, deleting *8* if we subsequently decide that the list should only contain numbers less than *6*.

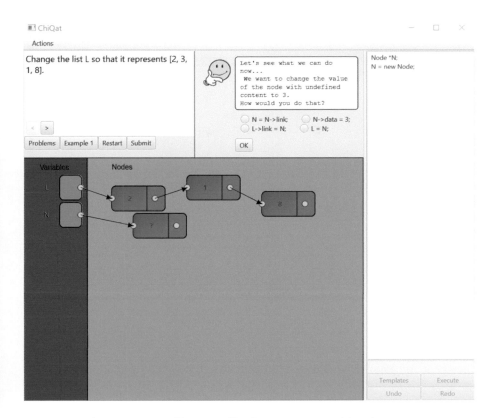

Figure A.1   A linked list (from a ChiQat-Tutor problem).

A.2   STACKS

Like lists, stacks are linear structures, in which the information is stored and accessed sequentially. A stack is characterized by the strict *policy* that has to be followed when storing and accessing the information. Imagine a stack in CS as a stack in real life, for example, a stack of trays in a cafeteria. Now imagine that you can only put a new tray on top of the existing stack, or only take the top tray from the stack; i.e., you are not allowed to search further down the stack to find a cleaner tray. That is exactly how a stack in CS conceptually works: you can only access the "top" element of the stack. Hence, *searching* for an element in a stack is not allowed, and inserting / deleting are limited to the only accessible position, which is technically called the *top* of the stack.

A.3   BINARY SEARCH TREES (BSTS)

Unlike lists and stacks, which are one-dimensional data structures, BSTs are two-dimensional structures, as shown in the bottom left pane of the ChiQat-Tutor screenshot shown in Figure A.2 (note that conventionally in

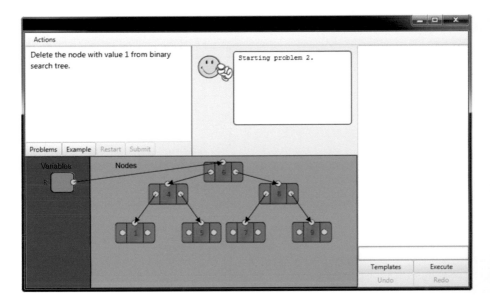

Figure A.2 A binary search tree.

CS, trees are shown as growing downwards from the root, as opposed to upward).

A typical node of a BST is composed of three parts: the data to be stored (i.e., the numbers in Figure A.2) and two pointers to two other nodes, called *left child* and *right child*. Hence, whereas a node in a list has only one successor, a node in a BST has two successors, one on the left and one on the right. If either child of a node does not exist, the corresponding pointer is set to *null* – see all the *leaves* at the third level of the tree[1] in Figure A.2 (in a CS tree, nodes without any successors are called *leaves*). Additionally, BSTs satisfy an ordering property: each node contains a value larger than any values contained in its left subtree (i.e., the tree whose root is the left child of the node in question), and smaller than any values in the right subtree (i.e., the tree whose root is the right child of the node in question). For example, in Figure A.2, the root node contains 6, which is larger than any values contained in the nodes to the left of 6, and smaller than the values contained in any node to the right of 6. The same applies if we descend lower in the tree, e.g., to the left of 6: 4 is larger than values contained in any node to its left, and smaller than values contained in any node to its right.

There are many operations that can be performed on a BST. The fundamental ones include *searching* for an element in the tree, which can be done very efficiently thanks to the ordering property of the structure; *inserting* a

---

[1] Technically, the leaves in Figure A.2 are at level two, not three, since the level of a node is the number of edges between it and the root.

new node in the tree; and *deleting* an existing node from the tree (as requested in the problem presented in the top left window in Figure A.2). Insertion and deletion are relatively complex algorithms, since the ordering property of the BST must be maintained.

Pre-/Post-Tests

CONTENTS

W E include here the pre-/post-tests used in the human–human tutoring data collection, and in ChiQat-Tutor. The pre-test and the post-test were identical in both the cases.

B.1   PRE-/POST-TEST FOR HUMAN TUTORING

You have the following two *linked lists*, starting from the head pointers H1 and H2. You also have a temporary pointer T.

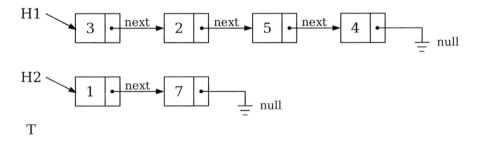

1. Look at the following procedure. The procedure is written in pseudo C/C++/Java, but don't worry about programming details such as declarations, etc. What is the status of the data structures after its execution? Draw a picture representing them.

```
T = H2;
while (T.next ≠ null) {
    T = T.next;
}
T.next = H1;
```

2. Consider the following "variation" of the same procedure. Why doesn't it work?

```
T = H2;
while (T ≠ null) {
    T = T.next;
}
T.next = H1;
```

You have a *stack* data structure. The operations defined on it are *push*, *pop* and *top*, with the usual semantics for stacks.

3. Your stack is initially empty. Write down the state of the stack and the content of the variable x after executing each of the following operations.

Operation	Stack	x
x = "A";		
push("B");		
push("C");		
push(x);		
pop();		
x = top();		
push("A");		
push(top());		

4. Mention an application (either a computer application or a real-world situation) in which you think a stack structure may be appropriate.

The following picture represents a *binary search tree* (BST).

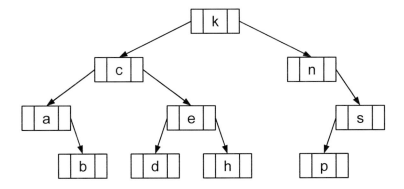

5. Show the keys with which "d" will be compared if you search for "d" in the given tree.

6. Insert a node containing "m" into the binary search tree. Draw the resulting tree.

7. Delete the node that contains "c" from the original binary search tree. Draw two possible final trees.

8. In general, is it more difficult to insert a new node in a BST or to delete an existing node from a BST? Explain why.

B.2 PRE-/POST-TEST FOR CHIQAT (LINKED LIST PROBLEMS)

The pre-post-test for linked lists, for the ChiQat-Tutor system, uses the same Problems 1 and 2 from the human tutoring pre/post-test, repeated here for convenience, and adds a third one.

You have the following two *linked lists*, starting from the head pointers H1 and H2. You also have a temporary pointer T.

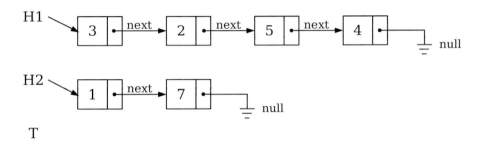

1. Look at the following procedure. The procedure is written in pseudo C/C++/Java, but don't worry about programming details such as declarations etc. What is the status of the data structures after its execution? Draw a picture representing them.

```
T = H2;
while (T.next != null) {
    T = T.next;
}
T.next = H1;
```

2. Consider the following "variation" of the same procedure. Why doesn't it work?

```
T = H2;
while (T != null) {
    T = T.next;
}
T.next = H1;
```

3. Write a sequence of operations, in pseudo-code or in a programming language of your choice, that moves the first node of the original list H1 to the end of the list.

Annotation Manuals

CONTENTS

WE provide the manuals we developed to annotate our human-human tutoring dialogues for elementary dialogue acts, student initiative, and episodic strategies, namely, worked out examples and analogies. Our coders had access to both audio and video files and were instructed to refer to both when coding the transcripts.

As concerns the numerous examples that follow, we include the session they come from, e.g., F05-12, and the line numbers from that session; T is for Tutor, and S for student. If an example is not marked with a session number, then it may not be an authentic example from the transcripts, but rather, it is an example similar to others in the data and was provided to facilitate coding.

As discussed in Chapter 3, our transcription conventions are a subset of those from CHAT, the transcription manual of the CHILDES project [272].[1] Transcription diacritics have the following meaning: '+...' marks trailing; +// self-interruptions; +^ a quick uptake, namely, cases where an utterance quickly follows the last utterance of the preceding speaker without the customary short pause between utterances; # a short pause, *um, *mmhm and similar, filled pauses;[2] angle brackets mark abandoned speech, or overlap with other materials: if they occur in two adjacent utterances by two different speakers, two additional markers are used in the two speakers' utterances, [>] and [<] respectively; 'xxx' marks unintelligible speech; a form like (be)cause marks a shortened form, i.e., the speaker actually said cause; [: text] represents using standard spelling for a spoken variant (e.g. [: going to] for gonna) and [^text] describes an event occurring at that point, most often some other sound. The diacritic @ is used to mark special symbols, in our case letters: for example, f@l signifies that the speaker named the letter "f" (in our data structures, node values are letters or numbers).

A note to the reader: What follows is taken verbatim from the coding manuals. The coding manuals were written at different points in time, and with different styles; the coding manuals for worked out examples and analogies were written years after the coding manual for dialogue acts, and the coders superimposed their annotations for worked out examples and analogies on the dialogues previously annotated for dialogue acts. Whereas for clarity we have occasionally rearranged categories or presentation of materials, or introduced formatting changes, we have been trying to be as faithful as possible to the originals, since coders coded our data according to the manuals as had been written at the time.

C.1 DIALOGUE ACT MANUAL

This section describes direct procedural instruction (DPI), direct declarative instruction (DDI), prompts and feedback.

C.1.1 Direct Procedural Instruction: DPI

In the context of problem solving, the category direct procedural instruction (DPI) should be applied when the tutor directly tells the student what to do. This label can be applied only to tutor's utterances. Do not apply it to student

---

[1] Available at https://talkbank.org/manuals/CHAT.pdf, latest update August 2020.
[2] Currently, CHAT marks filled pauses with &-um etc.

utterances. Always refer to the coding scheme when you encounter ambiguous utterances. ALWAYS watch the video. Some cases can only be disambiguated by listening to the prosody of the speakers, or by seeing what they are writing on their scratch paper.

1. Correct steps that lead to the solution of a problem should be marked as DPI (examples from F0512).

| T | 77 | It doesn't matter where you part point f@l xxx just point at that node. |

| T | 468 | and that is nothing there, so we put six right there. |
| T | 469 | and obviously as you said a minute ago, you put four right here. |

2. High-level steps or subgoals should be marked as DPI. Subgoals also include top-level goals. For example, the tutor may say, *here, they want us to delete the parent.* This is a DPI (examples from F05-12):

| T | 40 | it wants us to put the new node that contains g@l in it, after the node that contains b@l. |

| T | 87 | So let me show you how to find a node that has b@l in it. |

However, subgoals posed in the form of a question are NOT to be marked as DPI. For example, the tutor may say, *how do we do action X?*

Finally, some utterances may seem like subgoals but are indeed not relevant to the task. For example, the tutor may say, *let's look at this problem now.* These instances are NOT DPI.

3. Tactics and strategies should be marked as DPI (examples from F05-12):

| T | 41 | so with these kind of problems, the first thing I have to say is always draw pictures. |

| T | 474 | Well alright so that's why I want to do them in this order, search, and then if it fails, put it in there. |

4. When the tutor is repeating the same or a similar concept in different utterances, and both utterances are instances of DPI, then mark both of them (examples from F05-12):

```
T   41      So with these kind of problems, the first thing
            I have to say is always draw pictures.
T   42-46   [omitted]
T   47      So always draw pictures, particularly the after
            picture, which you want.
```

In the case of adjacent repeated utterances, that is, the tutor repeating the same or a similar concept one after the other, then the adjacent utterances are to be marked as DPI.

Sometimes, the tutor may include an explanation that follows the instruction. If the instruction and the subsequent explanation are in the same utterance then it is marked as DPI. In contrast, if the instruction and the subsequent explanation are in separate yet adjacent utterances, then only the instruction and NOT the explanation are to be marked as DPI (from F05-12):

```
(YES!)   T   248   So, the initialization is s@1 is null
                   (be)cause there's nothing behind the e@1.
```

5. In many occasions, the tutor talks in first person (*I do this... I have done that...*) or in third person (*This is what people do...*), but what the tutor really means is to provide direct instruction to the student (examples from F05-12):

```
T   525      so what people try to do is move two up,
             (be)cause two has two kids, and by the way two
             can't come up because <three and four can't be
             on the left side> [>]
```
```
T   611-612  um, so I'm pushing this value onto a stack.
T   612      so I'm pushing g@1 back on.
```

Sometimes, the tutor may provide "suggestive" DPI. For example, the tutor may say, *maybe you could do action X?* In such cases, the utterance is to be marked as DPI provided that the instruction is a correct procedure (from F05-12):

```
T   577   Or, we could have brought the four up, and once
          again the four wouldn't necessarily be a leaf if
          I put the three and a half right here xxx right
          here.
```

6. General domain knowledge that the tutor is telling the student (declarative instruction) should NOT be marked as DPI. Sometimes, instances of declarative instruction can be mistaken as procedural instruction. The

main difference is that elements of declarative instruction are usually not bound to a specific problem. In other words, declarative instruction tends to be more context-general and not context-specific. Furthermore, DPI is when the tutor tells the student what task to perform, whereas declarative instruction is when the tutor tells the student about context-general computer science knowledge.

An ambiguous utterance may be encountered when the tutor describes the problem or gives a description of the problem to the student. For example, the tutor may say, *now node m@l points to node k@l.* This utterance is NOT a DPI because the tutor is not directly instructing the student what task to perform. In other words, it is NOT a DPI when the tutor merely provides a description of the problem. However, there could be exceptions. For example, *the only thing you can do in this situation is X,* is ostensibly a descriptive statement, but it is also clearly a DPI, that is, to do X next.

Similarly, the tutor may describe the end result of an action. This is NOT to be marked as DPI. For example, the tutor may say, *Because we did X, node Y points to node Z.*

Furthermore, the tutor may be writing something relevant to the discussion. Here, there are two cases: the tutor may be writing out what actions the student wants to perform. This is NOT a DPI. The second case may involve the tutor writing out an instruction. This is DPI. For example, the tutor says, *do action X* and concurrently writes out the matching pseudo-code. Be careful to notice the context of the utterance in order to disambiguate mere description or declarative instruction versus DPI (all examples from F05-12):

(NO!)		38	yeah, well, so in link list you could put something in the middle real easy.

(NO!)	T	78	and then uh +// let me just write that down +// each node has two parts.
(NO!)	T	79	you call that the info, you call this the next field.

(NO!)	T	101	Right now t@l is pointing at this node.

(NO!)	T	337	so a binary tree, first of all okay link list is s@l one dimensional kind of data structure
	S	338	mmhm.
(NO!)	T	339	and a tree is two dimensional data structure.
	T	340	okay.
(NO!)	T	341	And the way you keep track of the tree is through a pointer to the first node which is called the root.

7. Sometimes, the tutor is creating an incorrect example or following an incorrect procedure on purpose, to make the student reason about possible incorrect scenarios. In these cases, the "wrong" steps should NOT be marked as instances of DPI (example from F05-12):

(NO!)	T	216-219	so this isn't a@l so we moved t@l over here we found a@l.
	T	220	do you see a problem?
(NO!)	T	221	I have found the node a@l, see here I found the node b@l, and then I put g@l in after it.
(NO!)	T	222	here I have found the node a@l and now the link I have to change is +...

The tutor may utter a DPI that is actually wrong but the tutor believes to be correct. In these instances, the utterance is to be marked as DPI. To err is human.

Sometimes, an incorrect procedure is difficult to determine especially for coders who have little or no background in computer science. In such cases, one should discuss the utterance in question with a coder who has substantial computer science experience.

8. Sometimes the tutor is using incomplete sentences. Mark them as DPI only if those incomplete sentences are instructing the student about what task to perform (example from F05-12):

(NO!)	T	265	now look, s@l's new value is +...
	T	266	you can't say s@l equals s@l dot is the next of s@l because s@l is null and doesn't have a next.
(NO!)	T	267	But there's a simple thing you can say, s@l's new value is +...

Sometimes, the tutor completes his or her own incomplete utterance.

For example, the tutor may say, *so if you want to delete node X then you should ... [short pause] ... do action Y.* In this example, the tutor leaves his or her utterance incomplete so the student can answer but the student does not answer thus, the tutor completes the utterance. This is DPI.

9. Sometimes, the tutor discusses hypothetical or *suppose you...* utterances. For example, if the tutor says, *suppose that so and so is the case, then we need to do X and Y.* This is DPI. However, hypothetical or *suppose you...* scenarios that are invented for the purpose of showing the consequences of errors are NOT a DPI (refer to item 7 above).

10. Negative *don't do X* utterances either are or are not DPI, depending on how tightly they constrain the learner's actions. *Don't do X,* in a situation or context in which that statement leaves completely open what to do instead, are NOT DPI; but, *don't do X,* in scenarios where NOT doing X clearly implies doing Y instead, are DPI. For example, *don't leave the door open* is DPI because the only way to follow the instruction is to close the door, so it tightly constrains the student's possible actions. *Don't paint the house red* is NOT a DPI, because it is very ambiguous, given that there are many other colors to paint with.

C.1.2   Direct Declarative Instruction: DDI

The tutor provides facts about the domain or a specific problem. This may be achieved by giving general information about data structures (e.g., properties of binary search trees), or by telling the student relevant information about a specific problem (e.g., the root node is five). The key here is that the tutor is telling the student something that he or she ostensibly does not already know.

C.1.2.1   *NOT DDI (NO!)*

1. DDI is not DPI. The two categories are mutually exclusive.

2. Common sense knowledge is not DDI (common sense knowledge is knowledge that is highly likely to be already known by the student) (from F05-01):

T   42   ten is greater than eight.
T   58   ten is less than eleven.

3. The utterance has nothing to do with a specific problem or the background of data structures (from F05-01):

T   122   it's just easier to draw a circle when you're used to drawing a square.

4. The utterance is too abstract or too far removed from directly addressing the problem (from F05-01):

```
T   182   there's a lot of +// there's a lot of work
          involved.
```

5. The utterance is describing what the tutor is or was doing (from F05-02):

```
T   7   xxx I was just finishing # this picture # so let me
        do that before I forget.
```

6. Questions asked by the tutor are not DDI.

C.1.2.2   DDI (YES!)

DDI utterances must be informative. DDI utterances must have some domain-relevant knowledge that is intended to be conveyed to the student.

1. DDI provides students with domain-general knowledge about binary search trees (BSTs), linked lists or stacks.

   Examples from F05-01:

```
T   88   the standard format is # right child is always
         greater than the parent # left child is always
         less than the parent.
```

```
T   100   so # um by convention the right child is always
          greater than its parent and the left child is
          always less than its parent.
```

```
T   266   the thought process of it is # you're # whatever
          you'll search for most often will always be a
          top.
```

   Examples from F05-02:

```
T   27   k@l [: okay] so an array is a list # but the
         things live consecutively and they live in memory
         together.
T   28   +^here they don't live together # so instead of #
         living together each one tells you where the next
         one is.
```

2. DDI is a conclusion or result of an action or its subsequent explanation Below are conclusions or results and their explanations (examples from F05-01).

| T | 143 | so now # since we've eliminated nine # it's gone. |
| T | 144 | nothing is pointing to nine. |

| T | 233 | so we did the minimal +// minimal amount of +// minimal amount of work that we had to do because now we didn't have to +// seven didn't have to go reestablish connections with anyone else. |

| T | 337 | when you did the search already you already found the appropriate place (be)cause that's what you did to find it. |

Below are conclusions or results.

Examples from F05-01:

T	63	so this is not a binary search tree.
T	64-79	[Omitted]
T	80	uh seven's left child is five which is less than seven.
T	81	[Omitted]
T	82	so this would be a binary search tree.
T	83-87	[Omitted]
T	88	it's not a typical binary search tree.

Example from F05-02:

| T | 442 | so you have failed # okay? |

3. Tutor's response to a student's question (from F05-01):

| S | 190 | all the way down? |
| T | 191 | everything would be on the left child # nothing would be on the right child. |

4. Utterances using "You will/are/have..." (Second-person DDI). Examples from F05-01:

T	255	so if you're not doing something so it would keep it balanced but at the cost of having to do several rotations so it's expensive.
T	256	[Omitted]
T	257	so if you have a huge tree and you do a couple inserts and you do rotations it might be rotated several times.

| T | 475 | so # you can see we have +// you'll get a different tree each time depending on the order of insertion. |

5. Utterances describing the tutoring materials (e.g., pictures). Examples from F05-02:

```
T   11      [^makes same clicking noise] so this is the
            information.
T   12      and then ## there is a link to the next node.
T   13      so this is a list with four nodes # [^makes
            same clicking noise] # obviously c@1 k@1 b@1
            f@1.
T   14-17   [Omitted]
T   18      so h@1 one is the header to the first list that
            tells you where the first node is.
```

6. Utterances using "can't". From F05-02:

```
T   36   but <you can't> [/] really just <## look at> [>]
         this and tell how many elements are on the list.
```

7. Solutions or potential solutions to a problem or sub-problem, which is not DPI. From F05-02:

```
T   550   right. so <the three> [/] the three is one of
          things we could promote.
```

8. Hypothetical scenarios (IF you did X, THEN Y). These may take the form of if-then statements. For example, *if you do X, then you'll end up with Y*. If such an utterance is informative and NOT DPI, then it is DDI. From F05-01:

```
T   175   so if you added six in # that's what we would
          get.
```
```
T   214   so now by making seven the new parent # we
          inherit all the fives.
```

9. DDI may be embedded within a positive or negative feedback episode. DDI may be a feedback response to a student's question.

```
FB+ START  PT    T 572   on this side three is the +...
                 S 573   bigger +...
FB+ END    DDI   T 574   the biggest.
```

C.1.3 Prompt

Tutor prompts occur when the tutor attempts to elicit an informative utterance from the student.

Exceptions:

1. A question which is a subclause in a declarative sentence is not a prompt (examples from F05-14, S06-42 and S06-43, respectively).

(NO!)	T	426	and g@1, g@1 is just like k@1, can I get, can I cut in front of you in line?
(NO!)	T	578	so we go ask Adam who is behind you?
(NO!)	T	514	we go see Carl, and we ask Carl, who is behind you?

2. Questions related to the tutoring session are not prompts (examples from F05-12 and S06-49 respectively)

	S	4	my name's Scott.
(NO!)	T	5	Scott?
(NO!)	T	499	now, what we're going to do is +// how much time do we have left?

C.1.3.1 Types of Prompts

Specific-Prompt[3]

Definition: Trying to get a specific response from the student.

Explanation: This type typically takes the form of *what, where, when, why, how,* and *yes* or *no* questions, like *What would the next letter be?*. Examples from F05-11:

(YES!)	T	79	that's not b@1 so what do we want to do?
(YES!)	T	102	well, where is that?
	T	103	that's the next field, we know that N@1 points to
(YES!)	T	104	so we'll say next of N@1 is equal to +// hmm how do I make it point here?

General Prompt

Definition: The tutor figures out what to do next.

Explanation: 1. Not a specific question, 2. Not related to any specific problem. Examples are *Why don't you try this problem?* or *Any questions?*

---

[3]The coding manual defines different types of prompts, but the data is only coded as Prompt, not for different types.

Diagnosing

Definition: Trying to determine what the student is doing or trying to determine the student's knowledge state.

Explanation: When the student answers a question and the tutor reveals that the student's answer is incorrect then, the tutor attempts to assess the situation. The sequence is listed below:

1. Tutor asks the student a question.

2. Tutor gives the student negative feedback.

3. Tutor tries to assess the student's incorrect answer.

Example from F05-09; note that the two Prompts in lines 637 and 638 are specific questions, while the one in 640 is of the *Diagnosing* type.

FB- START	PT	T 637	but could nine be the root of this tree?
	PT	T 638	with somehow this being on the left and that being on the right?
		S 639	it wouldn't be a problem.
	PT	T 640	well it would it still be a binary search tree?
FB- END	DDI	T 641	remember left is less and right is more, so if nine moved up,

Confirm-OK.

Definition. Questions which are used by the tutor to test if the student understands what he or she is talking about, or to see if the student is paying attention. For example, "OK?", "right?", "alright?". In what follows, the first two examples are from F05-10, the next three from F05-11, and the last one from F05-13.

T	537	so if I put the four in the tree, it would be here, right?
T	575	pop a stack gives you that, ok?
T	90	and of course now the info of t@l is b@l so we exit the loop, alright?
T	283	*uh, if we're looking for e@l, then this loop does nothing, alright?
T	414	ok, got that idea?
T	364	make sense?

General-OK.[4] *Definition.* These kinds of OK's are the tutor's responses to the student's responses.

---

[4]This type of prompt was defined, but the corresponding examples are left unlabeled in the coded data. We include it here for fidelity to the original coding manual.

Explanation. For example, the tutor says something, then the student responds to the tutor, and then the tutor says "OK?", "Right?", "Alright?". In what follows, the first example is from F05-11, the second from F05-12, and the third from F05-14.

T	308	um, so basically we changed a link but if we're actually deleting the first node, we change the header.
S	309	mmhm.
T	310	ok?

T	44	`<so we just change a couple of links here> [//]` actually change one link and add another link, so that's the after picture.
S	45	+< okay.
T	46	ok?

T	11	it starts at the bottom first.
S	12	mmhm.
T	13	ok

Fill in Blank

Definition The tutor starts an utterance but then does not complete it in the hopes of having the student complete the utterance.

Explanation There are two types of uncompleted utterances:

1. Tutor was actually expecting student to finish the utterance

2. Tutor was not expecting student to finish the utterance, but the student did so anyway.

Examples (from F05-11 and S06-55):

T	511	if you want the biggest over here, well, you got to go left to get into this tree and then you go +...
S	512	right until it's +...

T	333	yeah, so we +...
S	334	pop and push.

Procedure

Definition: The tutor assesses the student's knowledge state regarding basic data structures.

Examples (the first two from F05-11, the third from F05-12):

T	1	do you know what this is?

T	553	I also assume you have not heard of stack?

T	8	uh, do you know what a linked list is?

C.1.4 Feedback

To code for feedback, we defined *episodes* as a sequence of consecutive utterances that can be grouped together into positive or negative feedback.

C.1.4.1 Positive Feedback

The student says or does something correct, either spontaneously or after being prompted by the tutor. The tutor acknowledges the correctness of the student's claim, and possibly elaborates on it with further explanation. Example (from F05-12):

```
              T   217   do you see a problem?
              T   218   I have found the node a@l, see here I found
                        the node b@l, and then I put g@l in after
                        it.
  FB+_START   T   219   here I have found the node a@l and now the
                        link I have to change is +...
              S   220   ++ you have to link e@l <over xxx.> [>]
  FB+_END     T   221   [<] <yeah> I have to go back to this one.
              S   222   *mmhm
              T   223   so I *uh once I'm here, this key is here, I
                        can't go backwards.
```

C.1.4.2 Negative Feedback

The student says or does something wrong, either spontaneously or as an answer to a tutor's prompt. The tutor reacts to the mistake and possibly provides some form of explanation or remediation. Example from F05-12:

```
  FB-_START   S   107   <so you> [>] <you won't get the same> [//]
                        would you get the same point out of writing
                        t@l close to c@l at the top?
              T   108   oh, t@l equals c@l.
              T   109   no because you would have a type mismatch.
  FB-_END     T   110   t@l <is a pointer> [//] is an address, and
                        this is contents.
```

C.1.4.3 General Guidelines and Special Cases

1. When consecutive positive or negative feedback episodes are present in the data, it may happen that the end of an episode overlaps with the beginning of the following one. In this case, start the beginning of the new episode one utterance later. Example:

	S	460	+< six is xxx xx so you would have to move to the right.
FB+_END	T	461	it would have to be over there.
FB+_START	T	462	if the 6 was in this tree it would have to be +...
	S	463	++ to the left of nine.
FB+_END	T	464	and then it would have to be +...
FB+_START	S	465	++ left of eight.
FB+_END	T	466	and then it would have to be +...
FB+_START	S	467	++ left of seven
FB+_END	T	468	and that is nothing there, so we put six right there.

2. Sometimes, the tutor prompts the student, expecting an answer from him/her. In this case, include the tutor's prompt in the episode. In other cases, the tutor is just interrupted by the student, and the tutor provides feedback on what the student says. In this case, do *not* include the utterance of the tutor right before the interruption. Examples from F05-12:

FB+_START	T	547	or if you pick the smallest one over here +...
	S	548	it'll be greater than.
FB+_END	T	549	it'll be greater than five, since we entered all these guys automatically, and it's the smallest of these so it's less than these guys.

	T	534	<six would be ok because lets see> [//] ok six would be ok on that side +...
FB+_START	S	535	and to move this to this I would move four.
FB+_END	T	536	yeah exactly.

3. Sometimes, the participants of the conversation repeat things not to provide feedback, but just because they do not hear each other. If this happens at the beginning or at the end of a feedback episode, do not include the repetition in the episode. Example from F05-12:

	S	592	I also had a [?] top.
	T	593	I'm sorry?
FB+_START	S	594	you also had an a [?] top.
	T	595	yeah, okay.
FB+_END	T	596	right.

4. Sometimes, there is no real feedback. The tutor might say something that appears as feedback, but in fact he is ignoring what the student just said. Make sure *not* to mark such an episode as feedback. These cases might be difficult to detect. Watching the video helps in these situations. Example from F05-12:

```
T   281   it has to be +...
S   282   it has to be a step <before q@1.> [>]
T   283   [<] <it has to be before> we update it.
S   284   yeah
T   285   otherwise, t@1 would be s@1's new value.
T   286   so this is s@1 equals t@1, and s@1 is t@1's old
          value.
```

5. Mark only those episodes where the topic is relevant to the subject domain (in our case, linked lists, stacks and binary search trees). The following example can be considered feedback about social behavior, and should *not* be included in your annotation (from S06-48):

```
T   694   so now George is here in the line.
T   695   right?
T   696   okay.
S   697   that's very rude.
T   698   yes he's rude and *uh # Bob's no better for
          letting him behind him.
```

6. Pay particular attention to those words like "alright" and "okay." Such words might be part of the feedback, but sometimes, they are just used in the sense of "let's move on." In these cases, if they occur at the end of an episode, do not include them in the episode. Example from S06-57:

```
FB+_START   T   420   and d@1?
            S   421   these two should be reversed so +...
            T   422   yeah, in fact, the whole thing is
                      totally reversed.
            S   423   yeah.
            T   424   yeah, five and eleven are reversed,
                      and then ten is on the right, and
                      twelve is on the left, and four is on
                      the right, everything is flipped.
FB+_END     T   425   total flip.
            T   426   alright.
```

C.2   STUDENT INITIATIVE (SI)

SI occurs when the student proactively produces an utterance.

1. The student asks the tutor a question without being prompted by the tutor.

2. The student finishes the tutor's utterance which the tutor does not expect the student to complete.

3. The student disagrees with the tutor.

4. The student attempts to show the tutor that he or she understands the material.

Positive examples of SI in the following two excerpts from dialogue F05-09:

	T	440	*uh a binary tree is kind of like mother and father and xxx
(YES!)	S	441	a family tree.

	T	529	yeah we're looking for eight so we go right.
	T	530	and we find a nine.
(YES!)	S	531	so you got to go to the left.

Rules:

1. Separated utterances should all be tagged.

2. Utterances which answer a tutor's question are not SI.

3. Utterances which continue the previous initiative utterance should be tagged.

4. Some Student Initiatives (SI) may be several utterances away from a Tutor's prompt but are NOT SI if the utterance is related to the earlier tutor prompt. For example, a tutor may ask a student a question, but the student's answer to the question is located several lines down from the Tutor's question. This is NOT SI.

5. SI utterances must be informative and related to data structures.

6. Utterances where the tutor expects the student to complete his or her sentence are NOT SI.

In the following example from dialogue M05-02, line 109 answers the tutor's question in line 108, and line 111 continues that answer, so neither of them is an SI:

	T	108	yeah, what's the problem there?
(NO!)	S	109	the problem is you can't +// there's nothing referring back here.
	T	110	that's right # okay.
(NO!)	S	111	so unless you have like a doubly linked list t@l won't be able to do that.

C.3   WORKED-OUT EXAMPLES

Definition: A worked out example is a step-by-step demonstration of how to perform a task or how to solve a problem [76]. Worked-out examples consist of a problem formulation, solution steps and the final solution [197].

C.3.1   Coding Categories

Worked-out examples are to be primarily marked with a single coding group, *worked-out*. There will also be two subgroups, *level1-worked-out* and *level2-worked-out*, which will signify nested worked out examples.
The *worked-out* code group contains 3 codes:

- begin-worked-out: The beginning of a worked out example episode

- end-worked-out: The ending of a worked out example episode

- single-worked-out: If a worked out example starts and finishes in one turn, this code marks that turn as a worked out example.

Similarly, the *level1-worked-out* code group contains 3 codes:

- lvl1–begin-worked-out

- lvl1–end-worked-out

- lvl1–single-worked-out

The *level2-worked-out* code group contains 3 codes:

- lvl2-begin-worked-out

- lvl2-end-worked-out

- lvl2-single-worked-out

The codes for *level1-worked-out* and *level2-worked-out* serve the same goals as the codes for *worked-out*.
The three coding groups are to be used in a hierarchical structure. A *worked-out* example episode can contain several *level1-worked-out* example episodes, a *level1-worked-out* example episode can contain several *level2-worked-out* example episodes. A child episode must start and end

within its parent episode. A *level1-worked-out* and *level2-worked-out* example episode cannot exist without a **direct** parent episode. A single-turn episode (marked with single-worked-out and lvl1-single-worked-out) cannot have child episodes.

C.3.2   Marking Worked-Out Examples

C.3.2.1   Outline

A worked-out example must possess the following:

- Specific problem formulation
- Steps to solve the problem
- Problem solution/conclusion

A worked out example is not:

- A problem to be solved by the student (however, can lead to one if the student does not solve the problem and the tutor solves it for them)
- A procedure that has no initial problem formulation

C.3.2.2   Examples

A worked out example starts when the tutor introduces a specific problem. Each of the following examples are the start of a worked out example. Examples (the first two from F05-01, the third from F05-11):

WOE START	T	176	now if we were to delete five +...
WOE START	T	259	um there's one where you do a search on a tree say we search for +// let's come here and say we searched for nine.
WOE START	T	281	suppose we were trying to delete e@1.

Before a tutor gives a specific example, they may give a general description or goal for the specific example. Those descriptions do not count as the beginning of a worked out example. In the following episodes, the worked-out example starts from the second turn (all examples are from F05-11):

	T	362	so we got to invent a search algorithm.
WOE START	T	363	so the first one is search for +// let me cover it up for you +// search for eight.

	T	418	case zero, delete a node with zero kids, which is called by the way a leaf.
WOE START	T	419	so for example, suppose you're going to take six back out.

	T	447	alright, the case which at first seems extremely difficult is case two, two kids.
WOE START	T	448	so example they say here is to delete the five.

A worked-out example ends when the tutor draws the final solution for the first time. An example of this can be seen next, when the solution is given for a YES/NO style question (a prompt), that starts the WOE (from S06-58). For worked-out examples about inserting/deleting/searching for a node in a data structure, the worked-out example ended when the tutor achieved the original goal.

WOE START, PT	T	81	so we're going to go through this one here and determine whether +// these four here and determine whether they're binary search trees or not, okay?
	T	82-94	[Omitted]
WOE END	T	95	so this is a binary search tree.

The tutors may give explanations about why they took the approach they demonstrated (see first example below, lines 89-94, from F05-01), or generalize the specific example to a class of examples (see second example below, lines 433-443, from F05-11), or elaborate the problem. The rule about whether to include those turns into the worked-out example episode is whether those turns happened before the tutors draw the final solution for the first time.

WOE START	T	81	family for the last one.
	T	82-85	[Omitted]
	T	86	five +// five is less than eight so it's not equal.
SI	S	87	+< and that is less than eight?
WOE END	T	88	it's not a typical binary search tree.
DDI	T	89	the standard format is # right child is always greater than the parent # left child is always less than the parent.
	S	90	okay.
	T	91	okay.
DDI	T	92	so here we have +// we have exactly the opposite.
DDI	T	93	we have right child is less than the parent and left child is greater than the parent.
DDI	T	94	so the right is +// so right is greater # left is less than.
	T	95	so here we don't have a binary search tree.

WOE START, DPI	T	426	and the example here is, they say to delete nine. so here's what they +// let's see what they say.
DPI	T	427	delete the node that contains nine, moving as few nodes as possible, could draw a resulting tree and also keep it a binary search tree.
DPI	T	428	it doesn't say that specifically but they want you to keep it a binary search tree.
DPI PT	T	429	so do you see an easy way to delete nine and not mess up the whole tree very much at all?
	S	430	make five point to +...
	T	431	Ok great.
WOE END, DPI	T	432	make five point to eight, exactly.
DDI	T	433	and to talk about that in general, if we're in this case with one kid, doesn't really matter if it's a left kid or a right kid.
	T	[434-442]	[Omitted]
DDI	T	443	so that case is simple and as you see, it modified almost nothing in the tree.

During an example, there is the possibility of the episode becoming interactive. There are several scenarios where this may happen. First, the tutor could be interrupted by the student when they are presenting the example, giving the solution to the example (from F05-11):

WOE START	T	410	*uh, as a simple example *uh, use fractions.
FB+ START, PT	T	411	ok, seven +// where would seven and a half go?
	S	412	right, left, left, the right.
FB+ END, WOE END, DDI	T	413	so seven and a half would hang off here.

Another scenario is when the student asks the tutor a question. Questions could be asked for clarification of what has just been presented to them. Thirdly, it is also possible for the tutor to start quizzing the student during an example. This would allow the tutor to assess if the material is being absorbed by the student.

For the examples in which the student participated, if the student gave the answers directly (see previous and next examples – next example from F05-03), and the tutor did not further explain/correct/demonstrate the solution steps, then the episode is NOT a worked-out example.

```
FB+ START   T  297  how (a)bout this guy? #
            S  298  it looks good.
FB+ END     T  299  looks good.
```

For a worked-out example, the solution may be given by the student, in that case, the example did not end when the solution was given. The example ended after the tutor gave solution steps (see next example, from F05-03):

```
FB+ START, PT, WOE START  T  287  is this a binary search
                                   tree?
PT                        T  288  you see the violation of
                                   that property anywhere or
                                   does it hold anywhere?
                          T  289  right here.
                          T  290  right.
FB+ END                   T  291  so it is not correct?
                          S  292  yeah.
DDI                       T  293  so left is less 1@1 1@1.
DDI                       T  294  left is less.
DDI                       T  295  so this is wrong.
WOE END, DDI              T  296  it shouldn't be to the
                                   right.
```

C.4 ANALOGY CODING MANUAL

This section describes the definition of analogy and the corresponding coding category that we used to annotate our tutoring corpus.

C.4.1 Definition

Gentner defines analogies as *partial similarities between different situations that support further inferences* [36]. Specifically, analogy is a kind of similarity in which the same system of relations holds across different objects. Analogies thus capture parallels across different situations. The analogical process is described as follows [151, p.36]:

(1) *Retrieval*: given some current situation in working memory, a prior similar or analogous example may be retrieved from long

term memory; (2) *mapping*: given two cases in working memory, mapping consists of *aligning* their representational structures to derive the commonalities and *projecting inferences* from one analog to the other. Mapping is followed by (3) *evaluation* of the analogy and its inferences and often by (4) *abstraction* of the structure common to both analogies. A further process that may occur in the course of mapping is (5) *rerepresentation*: adaptation or of one or both representations to improve the match.

C.4.2 Analogous Terms

Analogous terms are the words used to describe an analogy that are not the terminology common to the primary domain. Examples:

- *person* is an analogous term - *We are going to tell person F@l ok you are going to point to person E@l*

- *in front of* is an analogous term -*If we think of each of these as a person in line, person c@l knows who's in front of him*

If a term can be used in both the analogy and the primary domain, the term is not considered to be an analogous term unless the instructor explicitly describes a mapping. Examples include:

- *point is not* an analogous term - *We are going to tell F@l ok you are going to point to E@l*

- *element* is an analogous term - *like a grocery list or whatever. uh so it has got a first element, second element, third element, fourth element and so on.*

C.4.3 Coding Category

Analogies are to be primarily marked with a single coding group, analogy. The analogy code group contains 3 codes:

- begin-analogy: The beginning of an analogy episode.

  - The beginning of the analogy is marked when either of the following occur
    * the prior similar or analogous example is first mentioned
    * the prior similar or analogous example is revisited after a prior analogy ending (new mapping).

- end-analogy: The end of an analogy episode

 – The end of the analogy is marked when either of the following occurs (in order of precedence)

* when the current session's problem is revisited without the analogous terms introduced in the analogy for more than ten (n>10) utterances [5]

* a new problem is introduced without the analogous terms introduced in the analogy for more than ten (n>10) utterances

* the prior similar or analogous example is last mentioned

- single-analogy:

  – If an analogy starts and finishes in one turn, this code marks that turn as an analogy

  – If the prior similar or analogous example is revisited in a single line

C.4.4 Marking Analogies

An analogy must possess the following:

- Citing of a prior similar or analogous example to a current topic

- Explanation of at least one parallel characteristic between the topic and example (mapping)

In addition to the required attributes, an analogy may additionally possess the following:

- Description of where the analogy holds and does not hold in comparison to the current topic

- Modification of the analogy to fit the evaluated shortcomings in comparison

C.4.4.1 Examples

Binary search tree is explained using family tree analogy (from F05-09):

ANALOGY START	T	440	*uh a binary tree is kind of like mother and father and xxx
	S	441	a family tree.
	T	442	no that's not bad *uh that's bad.
	T	443	it's +// because families can have more than two kids.
	T	444	so here what it means is that binary is that each node can have two trees, two children.
ANALOGY END	T	445	+< two kids.

---

[5]end-analogy should be marked at the location where the count of utterances first began

Linked Lists are explained using a grocery list analogy (from F05-04):

ANALOGY START	T	10	like a grocery list or whatever.
ANALOGY END	T	11	uh so it has got a first element, second element, third element, fourth element and so on.

Stacks are explained using a stack of legos analogy (from F05-06):

ANALOGY START	T	219		think of the stack as a bunch of legos, okay?
	T	220		and each time you put out a lego +...
	T	221		okay <we'll call this>[//] we'll just go a@l b@l c@l d@l e@l, and so forth.
	T	222		okay?
	T	223		so we're stacking our legos up.
	T	224		if we want to take a lego off we can only take the lego off that we just inserted.
	T	225		right?
	T	226		(be)cause we're building from the bottom up.
	T	227		okay?
	T	228		so we can only <take in>[//] take off whatever we put in last.
	S	229		okay.
		[230–238]		Omitted
ANALOGY END	T	239		it's kind of like a stack of legos you can't just pull off the center piece.

Linked List Problem Set

CONTENTS

This appendix includes the seven problems in the linked list module of ChiQat-Tutor, and the system solution for each.

D.1 PROBLEM 1

Change the list L so that it represents [2, 3, 1, 8].

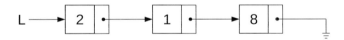

Solution:

```
Node t;
t = new Node();
t.data = 3;
t.link = L.link;
L.link = t;
```

D.2   PROBLEM 2

Change the list L1 so that it represents the concatenation of L1 and L2, i.e.,
[2, 9, 8, 3, 5, 1, 2].

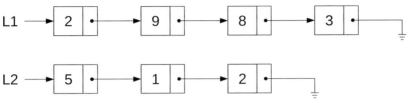

Solution:

```
Node T;
T = L1;
T = T.link;
T = T.link;
T = T.link;
T.link = L2;
```

D.3   PROBLEM 3

Delete the first node from list L.

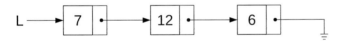

Solution:

```
Node t;
t = L.link;
delete L;
L = t;
```

D.4 PROBLEM 4

Delete the node from L that is pointed to by P.

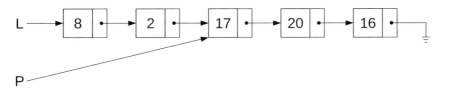

Solution:

```
Node t;
t = L;
t = t.link;
t.link = P.link;
delete P;
```

D.5 PROBLEM 5

Split the list L into two pieces: L should be the list of everything before the node with data value 2, R should be the list of everything including and after that node. So, when you are done, L = [12, 8, 15, 5], R = [2, 4, 17].

Solution:

```
Node t;
t = L;
t = t.link;
t = t.link;
t = t.link;
R = t.link;
t.link = null;
```

D.6   PROBLEM 6

Write a code fragment that will set pointer T to point to the node immediately preceding the node pointed to by pointer P in the list represented by L. The same code fragment should work in both scenarios depicted.

Scenario 1

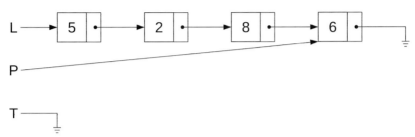

Scenario 2

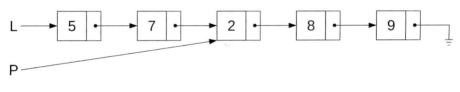

Solution:

```
T = L;
while (T.link != P) {
  T = T.link;
}
```

D.7 PROBLEM 7

When you write code to find a node in a list, you want the answer to be a pointer to the node you are looking for. If the node you are searching does not exist, the answer is simply NULL. Write a code fragment that finds the node containing 8 in list L by setting pointer P to the node, or NULL if the node can't be found. Your code should work for each of the given scenarios.

Scenario 1

P

Scenario 2

P

Scenario 3

P

Solution:

```
P = L;
while (P != null && P.data != 8) {
  P = P.next;
}
```

Stack Plugin Full Code

CONTENTS

This appendix includes the full code of the new Stack Plugin discussed in Chapter 7. This plugin is composed of five Java files and one CSS (Cascading Style Sheet) file:

1. PluginInstance.java

2. StackView.java

3. StackProblem.java

4. StackProblemStep.java

5. StackProblemFeedback.java

6. StackStyle.css

E.1 PLUGININSTANCE.JAVA

```
1  package edu.uic.cs.nlp.chiqat.plugin.stackplugin;
2
3  import edu.uic.cs.nlp.chiqat.framework.AppDataStore;
4  import edu.uic.cs.nlp.chiqat.framework.BluePrintNode;
5  import edu.uic.cs.nlp.chiqat.framework.DefaultMessageReceiver;
6  import edu.uic.cs.nlp.chiqat.framework.IMessageReceiver;
7  import edu.uic.cs.nlp.chiqat.framework.KeyValuePair;
8  import edu.uic.cs.nlp.chiqat.framework.LocalPluginConnector;
9  import edu.uic.cs.nlp.chiqat.framework.Message;
```

```
10  import edu.uic.cs.nlp.chiqat.framework.PluginConnector;
11  import edu.uic.cs.nlp.chiqat.plugin.base.BasePluginInstance;
12  import edu.uic.cs.nlp.chiqat.plugin.base.PluginViewDefinition;
13  import java.util.List;
14
15  public class PluginInstance extends BasePluginInstance implements
        IMessageReceiver {
16
17    public PluginInstance() {}
18
19    // Initalises the plugin
20    @Override
21    public boolean Init(
22        PluginConnector connector,
23        List<KeyValuePair> args,
24        BluePrintNode node,
25        AppDataStore appDataStore) {
26      super.Init(
27          connector,
28          args,
29          node,
30          appDataStore,
31          new DefaultMessageReceiver((LocalPluginConnector) connector,
              this));
32
33      // Create a view
34      PluginViewDefinition view = new PluginViewDefinition(this);
35      view.m_ViewType = StackView.class;
36      view.m_ViewTypeId = "StackView";
37      m_ViewDefs.add(view);
38
39      m_Receiver.start();
40
41      return true;
42    }
43
44    @Override
45    public boolean Destroy() {
46      super.Destroy();
47      if (m_Receiver != null) {
48        m_Receiver.Shutdown();
49      }
50      return true;
51    }
52
53    @Override
54    public void OnMessage(Message msg) {
55      super.OnMessage(msg);
56    }
```

```
57  }
```

E.2 STACKVIEW.JAVA

```
1   package edu.uic.cs.nlp.chiqat.plugin.stackplugin;
2
3   import edu.uic.cs.nlp.chiqat.plugin.base.PluginViewFx;
4   import javafx.scene.control.Button;
5   import javafx.scene.control.Label;
6   import javafx.scene.control.ScrollPane;
7   import javafx.scene.control.Spinner;
8   import javafx.scene.control.TextField;
9   import javafx.scene.control.TextFormatter;
10  import javafx.scene.layout.BorderPane;
11  import javafx.scene.layout.GridPane;
12  import javafx.scene.layout.HBox;
13  import javafx.scene.layout.Pane;
14  import javafx.scene.layout.VBox;
15  import javafx.util.StringConverter;
16  import javafx.util.converter.DefaultStringConverter;
17
18  /**
19   * User interface for stack tracing problems.
20   *
21   * @author Davide Fossati
22   */
23  public class StackView extends PluginViewFx {
24
25      public static final String STACK_TUTOR_CSS =
26          StackTutor.class.getResource("StackStyle.css").toExternalForm();
27
28      public static final String[] ALPHABET = {"A", "B", "C", "D", "E"};
29
30      public static final String[] VAR_NAMES = {"X", "Y", "A", "B", "C",
            "D", "E", "V"};
31
32      StackProblem problem;
33
34      TextField tfProblemLength;
35
36      TextField tfNumVars;
37
38      Spinner<Integer> spnProblemLength;
39
40      Spinner<Integer> spnNumVars;
41
42      GridPane problemGrid;
```

```
43
44    TextField[] tfStack;
45
46    TextField[][] tfVars;
47
48    Label lblFeedback;
49
50    Button btnCheckSolution;
51
52    Button btnShowSolution;
53
54    @Override
55    public boolean Init() {
56      Pane root = createMainPane();
57      root.getStylesheets().add(STACK_TUTOR_CSS);
58      root.getStyleClass().add("root");
59      this.getChildren().add(root);
60
61      // Make the root pane stretch when its container is resized
62      setTopAnchor(root, 0.0);
63      setBottomAnchor(root, 0.0);
64      setLeftAnchor(root, 0.0);
65      setRightAnchor(root, 0.0);
66
67      return true;
68    }
69
70    private BorderPane createMainPane() {
71      Label lblProblemLength = new Label("Number of Steps ");
72      spnProblemLength = new Spinner<Integer>(1, 1000, 8);
73
74      Label lblNumVars = new Label("Number of Variables ");
75      spnNumVars = new Spinner<Integer>(0, 100, 1);
76
77      Button btnNewProblem = new Button("New Problem");
78      btnNewProblem.setOnAction(e -> createProblem());
79
80      GridPane topArea = new GridPane();
81      topArea.add(lblProblemLength, 0, 0);
82      topArea.add(spnProblemLength, 1, 0);
83      topArea.add(lblNumVars, 0, 1);
84      topArea.add(spnNumVars, 1, 1);
85      topArea.add(btnNewProblem, 0, 2);
86
87      problemGrid = new GridPane();
88      problemGrid.getStyleClass().add("grid");
89
90      ScrollPane scroll = new ScrollPane();
91      scroll.setContent(problemGrid);
```

```
92
93      btnCheckSolution = new Button("Check Your Solution");
94      btnCheckSolution.setOnAction(e -> checkSolution());
95      btnCheckSolution.setDisable(true);
96
97      btnShowSolution = new Button("Show Correct Solution");
98      btnShowSolution.setOnAction(e -> showSolution());
99      btnShowSolution.setDisable(true);
100
101     lblFeedback = new Label();
102
103     HBox bottomButtons = new HBox(btnCheckSolution, btnShowSolution);
104     VBox bottomArea = new VBox(lblFeedback, bottomButtons);
105
106     BorderPane main = new BorderPane();
107     main.setTop(topArea);
108     main.setCenter(scroll);
109     main.setBottom(bottomArea);
110
111     return main;
112   }
113
114   private void createProblem() {
115     int numSteps = spnProblemLength.getValue();
116     String[] varNames = new String[spnNumVars.getValue()];
117     for (int i = 0; i < varNames.length; i++) {
118       if (i < VAR_NAMES.length - 1) {
119         varNames[i] = VAR_NAMES[i];
120       } else {
121         varNames[i] = VAR_NAMES[VAR_NAMES.length - 1] + (i + 1);
122       }
123     }
124
125     problem = new StackProblem(numSteps, ALPHABET, varNames);
126
127     problemGrid.getChildren().clear();
128     lblFeedback.setText("");
129
130     String[] sHead = new String[3 + problem.varNames.length];
131     sHead[0] = "Step";
132     sHead[1] = "Operation";
133     sHead[2] = "Stack";
134     for (int i = 0; i < problem.varNames.length; i++) {
135       sHead[i + 3] = problem.varNames[i];
136     }
137     Label[] head = new Label[sHead.length];
138     for (int i = 0; i < head.length; i++) {
139       head[i] = new Label(sHead[i]);
140       head[i].getStyleClass().add("problem-head");
```

```
141      problemGrid.add(head[i], i, 0);
142    }
143
144    tfStack = new TextField[problem.steps.length];
145    tfVars = new
              TextField[problem.varNames.length][problem.steps.length];
146    StringConverter<String> spaceRemover =
147         new DefaultStringConverter() {
148           @Override
149           public String toString(String s) {
150             return s == null ? "" : s.replaceAll("\\s", "");
151           }
152         };
153
154    for (int i = 0; i < problem.steps.length; i++) {
155      Label lblStepNumber = new Label("" + (i + 1) + ".");
156      Label lblOperation = new Label(problem.steps[i].operation);
157      lblOperation.getStyleClass().add("operation-text");
158      tfStack[i] = new TextField();
159      tfStack[i].getStyleClass().add("stack-field");
160      tfStack[i].setTextFormatter(new
                TextFormatter<String>(spaceRemover));
161
162      problemGrid.add(lblStepNumber, 0, i + 1);
163      problemGrid.add(lblOperation, 1, i + 1);
164      problemGrid.add(tfStack[i], 2, i + 1);
165
166      for (int v = 0; v < problem.varNames.length; v++) {
167        tfVars[v][i] = new TextField();
168        tfVars[v][i].getStyleClass().add("variable-field");
169        tfVars[v][i].setTextFormatter(new
                  TextFormatter<String>(spaceRemover));
170
171        problemGrid.add(tfVars[v][i], v + 3, i + 1);
172      }
173    }
174
175    btnCheckSolution.setDisable(false);
176    btnShowSolution.setDisable(true);
177  }
178
179  private void checkSolution() {
180    if (problem == null) {
181      return;
182    }
183
184    String[][] studentSolution = new String[problem.steps.length][1 +
              problem.varNames.length];
185
```

```
186      for (int i = 0; i < problem.steps.length; i++) {
187        studentSolution[i][0] = tfStack[i].getText();
188
189        for (int v = 0; v < problem.varNames.length; v++) {
190          studentSolution[i][v + 1] = tfVars[v][i].getText();
191        }
192      }
193
194      StackProblemFeedback feedback =
            problem.checkSolution(studentSolution);
195
196      for (int i = 0; i < problem.steps.length; i++) {
197        if (feedback.correctItems[i][0] == true) {
198          tfStack[i].setStyle("-fx-background-color: green");
199        } else {
200          tfStack[i].setStyle("-fx-background-color: red");
201        }
202
203        for (int v = 0; v < problem.varNames.length; v++) {
204          if (feedback.correctItems[i][v + 1] == true) {
205            tfVars[v][i].setStyle("-fx-background-color: green");
206          } else {
207            tfVars[v][i].setStyle("-fx-background-color: red");
208          }
209        }
210      }
211
212      lblFeedback.setText(feedback.getFeedback());
213
214      btnCheckSolution.setDisable(false);
215      btnShowSolution.setDisable(false);
216    }
217
218    private void showSolution() {
219      if (problem == null) {
220        return;
221      }
222
223      for (int i = 0; i < problem.steps.length; i++) {
224        tfStack[i].setText(problem.steps[i].stack);
225        tfStack[i].setEditable(false);
226        tfStack[i].setStyle(null);
227
228        for (int v = 0; v < problem.varNames.length; v++) {
229          tfVars[v][i].setText(problem.steps[i].vars[v]);
230          tfVars[v][i].setEditable(true);
231          tfVars[v][i].setStyle(null);
232        }
233      }
```

```
234
235       btnCheckSolution.setDisable(true);
236       btnShowSolution.setDisable(false);
237    }
238 }
```

E.3   STACKPROBLEM.JAVA

```
1  package edu.uic.cs.nlp.chiqat.plugin.stackplugin;
2  /**
3   * This class represents a stack tracing problem. A problem is
       composed of a stack data structure, a
4   * set of named variables, and a sequence of operations such as
       push(), pop(), top(), and variable
5   * assignments. The stack and the variables are initially empty.
6   *
7   * The goal of the problem is to predict the content of the stack and
       the variables after the
8   * execution of each operation in the sequence.
9   *
10  * @author Davide Fossati
11  */
12 public class StackProblem {
13
14   StackProblemStep[] steps; // sequence of steps composing the current
         problem
15   // each step is composed of an operation and the state of the memory
16   // after the execution of the operation
17
18   String[] alphabet; // symbols that can be inserted in the stack
         and/or assigned to variables
19   // each symbol should be one character long
20
21   String[] varNames; // names of the variables in memory (excluding
         the stack)
22
23   /**
24    * Constructor. Generates a random problem with a specified length
         and complexity.
25    *
26    * @param numSteps Number of steps of the problem. The actual number
         of steps constructed will be
27    *    at least the number of variables in the problem.
28    * @param alphabet Symbols that can be inserted in the stack and/or
         assigned to variables. Each
29    *    symbol should be one character long.
```

```
30     * @param varNames Names of the variables in memory (excluding the
               stack).
31     */
32    public StackProblem(int numSteps, final String[] alphabet, final
           String[] varNames) {
33      steps = new StackProblemStep[numSteps];
34      this.alphabet = alphabet;
35      this.varNames = varNames;
36
37      String stack = ""; // current content of the stack
38      String[] vars = new String[varNames.length]; // current content of
               the variables
39      for (int i = 0; i < vars.length; i++) {
40        vars[i] = "";
41      }
42
43      // The first few steps are variable initializations
44      for (int i = 0; i < varNames.length && i < numSteps; i++) {
45        steps[i] = new StackProblemStep();
46        vars[i] = randomLetter();
47        steps[i].operation = varNames[i] + " = \"" + vars[i] + "\"";
48        steps[i].stack = stack;
49        steps[i].vars = vars.clone();
50      }
51
52      // The remaining steps can be push(), pop(), top(), or variable
               assignments
53      for (int i = varNames.length; i < numSteps; i++) {
54        steps[i] = new StackProblemStep();
55        // Generate a random operation.
56        // If the stack is empty generate a push(),
57        // otherwise the probability of a push() operation is 35%
58        double p = Math.random();
59        if (stack.length() == 0 || p < 0.35) {
60          // Generate a push() operation.
61          // If the stack is empty we can only push a constant or a
                   variable,
62          // otherwise we can also push the top of the stack
63          if (stack.length() == 0 || Math.random() < 0.66) {
64            if (varNames.length > 0 && Math.random() < 0.5) {
65              // Push the content of a random variable.
66              // We assume all the variables have already been initialized
67              int v = (int) (Math.random() * varNames.length);
68              steps[i].operation = "push(" + varNames[v] + ")";
69              stack = push(stack, vars[v]);
70            } else {
71              // Push a random constant
72              String letter = randomLetter();
73              steps[i].operation = "push(\"" + letter + "\")";
```

```
74        stack = push(stack, letter);
75      }
76    } else {
77      // Push the result of a top() operation
78      String letter = top(stack);
79      steps[i].operation = "push(top())";
80      stack = push(stack, letter);
81    }
82
83  } else if (varNames.length > 0 && p < 0.70) {
84    // Generate a variable assignment.
85    // Pick a random variable
86    int v = (int) (Math.random() * varNames.length);
87    // Pick either a random constant, a top() operation, or a pop()
           operation
88    double p2 = Math.random();
89    if (p2 < 0.33) {
90      // Random constant
91      vars[v] = randomLetter();
92      steps[i].operation = varNames[v] + " = \"" + vars[v] + "\"";
93    } else if (p2 < 0.66) {
94      // top() operation
95      vars[v] = top(stack);
96      steps[i].operation = varNames[v] + " = top()";
97    } else {
98      // pop() operation
99      vars[v] = top(stack);
100     stack = pop(stack);
101     steps[i].operation = varNames[v] + " = pop()";
102   }
103
104  } else {
105    // Generate a pop() operation
106    steps[i].operation = "pop()";
107    stack = pop(stack);
108  }
109
110  // Update the final state of the stack and the variables in the
         new step
111  steps[i].stack = stack;
112  steps[i].vars = vars.clone();
113  }
114 }
115
116 /**
117  * Picks a random symbol from the current alphabet.
118  *
119  * @return A random String picked from the current alphabet.
120  */
```

```
121    private String randomLetter() {
122      int choice = (int) (Math.random() * alphabet.length);
123      return alphabet[choice];
124    }
125
126    /**
127     * Returns the stack with the new element pushed on top. The top of
                the stack is the last
128     * character of the String.
129     *
130     * @param stack String representation of the stack. Each element of
                the stack is a single
131     *     character with no separating spaces or other characters in
                between. The top of the stack is
132     *     the last character of the String.
133     * @param element The element to be pushed onto the stack. It should
                be a String with a single
134     *     character.
135     * @return The stack with the new element pushed on top.
136     */
137    private String push(String stack, String element) {
138      return stack + element;
139    }
140
141    /**
142     * Returns the stack after removing the top element. Unlike the
                traditional pop() operation for
143     * stacks, this implementation does NOT return the element that was
                removed. If the stack is
144     * empty, returns the empty String.
145     *
146     * @param stack String representation of the stack. Each element of
                the stack is a single
147     *     character with no separating spaces or other characters in
                between. The top of the stack is
148     *     the last character of the String.
149     * @return The stack after removing the top element, or the empty
                String if the stack was empty.
150     */
151    private String pop(String stack) {
152      if (stack != null && stack.length() > 0) {
153        return stack.substring(0, stack.length() - 1);
154      } else {
155        return "";
156      }
157    }
158
159    /**
160     * Returns the element at the top of the stack.
```

```
161    *
162    * @param stack String representation of the stack. Each element of
         the stack is a single
163    *   character with no separating spaces or other characters in
         between. The top of the stack is
164    *   the last character of the String.
165    * @return The element at the top of the stack, which is a String
         with a single character.
166    */
167   private String top(String stack) {
168     if (stack != null && stack.length() > 0) {
169       return stack.substring(stack.length() - 1);
170     } else {
171       return "";
172     }
173   }
174
175   /**
176    * Compares a given student's solution to the correct solution of
         this problem. Returns a
177    * StackProblemFeedback object containing: 1) a matrix indicating
         which items are correct; 2)
178    * verbal feedback for the student.
179    *
180    * @param studentSolution a matrix of string containing the
         student's solution. Each row
181    *   represents a step of the problem. The first column represents
         the content of the stack. The
182    *   following columns contain the values of the variables.
183    * @return A StackProblemFeedback object containing the result of
         the comparison.
184    */
185   public StackProblemFeedback checkSolution(String[][]
         studentSolution) {
186     boolean[][] correctItems = new boolean[steps.length][1 +
         varNames.length];
187
188     for (int i = 0; i < steps.length; i++) {
189       correctItems[i][0] =
             studentSolution[i][0].equals(steps[i].stack);
190
191       for (int v = 0; v < varNames.length; v++) {
192         correctItems[i][v + 1] = studentSolution[i][v +
             1].equals(steps[i].vars[v]);
193       }
194     }
195
196     return new StackProblemFeedback(correctItems);
197   }
```

```
498  }
```

E.4   STACKPROBLEMSTEP.JAVA

```
1   package edu.uic.cs.nlp.chiqat.plugin.stackplugin;
2   /**
3   * This class represents a single step for a stack tracing problem. A
        step contains an operation and
4   * the state of the memory (stack and variables) after the execution
        of that operation.
5   *
6   * @author Davide Fossati
7   */
8   class StackProblemStep {
9
10    String operation; // operation applied to the previous state of the
          stack/variables
11
12    String stack; // content of the stack after the execution of the
          operation
13
14    String[] vars; // content of the variables after the execution of
          the operation
15  }
```

E.5   STACKPROBLEMFEEDBACK.JAVA

```
1   package edu.uic.cs.nlp.chiqat.plugin.stackplugin;
2   /**
3   * This class represents the result of the comparison between a
        submitted solution and the correct
4   * solution for a stack tracing problem. This information is used to
        generate visual and verbal
5   * feedback for the student.
6   *
7   * @author Davide Fossati
8   */
9   public class StackProblemFeedback {
10
11    boolean[][] correctItems; // each element of this matrix is set to
          true if the corresponding item
12    // (stack or variable) in the original StackProblem submission was
          correct,
13    // false otherwise.
```

```
14
15    private int numCorrect; // number of correct items in the matrix
16
17    /**
18     * Constructor
19     *
20     * @param correctItems A Boolean matrix with the same size as the
               input items in the original
21     *    StackProblem submission. A value of true indicates that the
             submitted item matches the
22     *    correct solution.
23     */
24    public StackProblemFeedback(boolean[][] correctItems) {
25      this.correctItems = correctItems;
26
27      // count the number of correct items
28      numCorrect = 0;
29      for (int i = 0; i < correctItems.length; i++) {
30        for (int j = 0; j < correctItems[i].length; j++) {
31          if (correctItems[i][j] == true) {
32            numCorrect++;
33          }
34        }
35      }
36    }
37
38    /**
39     * Computes and returns verbal feedback based on the number of
             correct items in the student's
40     * solution.
41     *
42     * @return A String containing feedback for the student.
43     */
44    public String getFeedback() {
45      int numItems = correctItems.length * correctItems[0].length;
46      return String.format(
47          "Correct: %d/%d (%.1f%%)", numCorrect, numItems, 100.0 *
               numCorrect / numItems);
48    }
49  }
```

E.6   STACKSTYLE.CSS

```
1  .root {
2    -fx-font-size: 11pt;
3    -fx-padding: 20;
4  }
```

```
 5
 6  .grid {
 7     -fx-hgap: 10;
 8     -fx-vgap: 10;
 9     -fx-padding: 20;
10  }
11
12  .problem-head {
13     -fx-font-weight: bold;
14  }
15
16  .operation-text {
17     -fx-font-family: "Monospace";
18  }
19
20  .stack-field {
21     -fx-font-family: "Monospace";
22  }
23
24  .variable-field {
25     -fx-font-family: "Monospace";
26     -fx-pref-width: 70;
27  }
```

Bibliography

[1] AA.VV. Computing Curricula 2001 – Computer Science. Association for Computing Machinery, and IEEE Computer Society, December 2001. Report of the Joint Task Force.

[2] AA.VV. Subject benchmark statement: Computing. The Quality Assurance Agency for Higher Education, United Kingdom, 2007.

[3] AA.VV. Computer Science Curriculum 2013. Association for Computing Machinery, and IEEE Computer Society, December 2013. Report of the Joint Task Force.

[4] Shazia Afzal, Tejas Dhamecha, Nirmal Mukhi, Renuka Sindhgatta, Smit Marvaniya, Matthew Ventura, and Jessica Yarbro. Development and deployment of a large-scale dialog-based Intelligent Tutoring System. In *Proceedings of the 2019 Conference of the North American Chapter of the Association for Computational Linguistics: Human Language Technologies*, volume 2, pages 114–121. Association for Computational Linguistics, Minneapolis, MN, USA, 2019.

[5] Ali Ahmadvand, Jason Ingyu Choi, and Eugene Agichtein. Contextual dialogue act classification for open-domain conversational agents. In *Proceedings of the 42nd International ACM SIGIR Conference on Research and Development in Information Retrieval*, SIGIR'19, pages 1273–1276, New York, NY, USA, 2019. ACM.

[6] Hua Ai, Diane J. Litman, Kate Forbes-Riley, Mihai Rotaru, Joel Tetreault, and Amruta Purandare. Using system and user performance features to improve emotion detection in spoken tutoring dialogs. In *Ninth International Conference on Spoken Language Processing (INTERSPEECH 2006)*, pages 797–800, Pittsburgh, PA, Sep. 2006.

[7] Patricia Albacete, Pamela W. Jordan, Sandra Katz, Irene-Angelica Chounta, and Bruce M. McLaren. The impact of student model updates on contingent scaffolding in a natural-language tutoring system. In Seiji Isotani, Eva Millán, Amy Ogan, Peter Hastings, Bruce McLaren, and Rose Luckin, editors, *20th International Conference on Artificial Intelligence in Education (AIED 2019)*, pages 37–47, Chicago, IL, 2019. Springer International Publishing.

[8] Vincent Aleven and Kenneth R. Koedinger. An effective metacognitive strategy: learning by doing and explaining with a computer-based cognitive tutor. *Cognitive Science*, 26:147–179, 2002.

[9] Tahani Aljohani and Alexandra I Cristea. Predicting Learners' Demographics Characteristics: Deep Learning Ensemble Architecture for Learners' Characteristics Prediction in MOOCs. In *Proceedings of the 2019 4th International Conference on Information and Education Innovations*, Durham, United Kingdom, pages 23–27, 2019.

[10] Ali Alkhatlan and Jugal Kalita. Intelligent tutoring systems: A comprehensive historical survey with recent developments. *International Journal of Computer Applications*, 181(43), March 2019.

[11] James F. Allen, Bradford W. Miller, Eric K. Ringger, and Teresa Sikorski. A robust system for natural spoken dialogue. In *Proceedings of the 34th Annual Meeting of the Association for Computational Linguistics*, pages 62–70. Association for Computational Linguistics, Santa Cruz, CA, USA 1996.

[12] Zeyad Alshaikh, Lasang Tamang, and Vasile Rus. Experiments with a Socratic Intelligent Tutoring System for Source Code Understanding. In *The Thirty-Third International FLAIRS Conference*, Online, pages 457–460, 2020.

[13] John R. Anderson. *Language, Memory, and Thought*. Lawrence Erlbaum Associates, Hillsdale, NJ, 1976.

[14] John R. Anderson. *The Architecture of Cognition*. Harvard University Press, Cambridge, MA, 1983.

[15] John R. Anderson. Knowledge compilation: The general learning mechanism. In R. S. Michalski, J. G. Carbonell, and T. M. Mitchell, editors, *Machine Learning*, volume 5, pages 289–310. Kaufmann, Los Altos, CA, 1986.

[16] John R. Anderson, Frederick G. Conrad, and Albert T. Corbett. Skill acquisition and the LISP tutor. *Cognitive Science*, 13(4):467–505, 1989.

[17] John R. Anderson, Albert T. Corbett, Kenneth R. Koedinger, and R. Pelletier. Cognitive tutors: Lessons learned. *Journal of the Learning Sciences*, 4(2):167–207, 1995.

[18] Richard B. Anderson and Ryan D. Tweney. Artifactual power curves in forgetting. *Memory & Cognition*, 25(5):724–730, 1997.

[19] Giovanni Andreani, Giuseppe Di Fabbrizio, Mazin Gilbert, Daniel Gillick, Dilek Hakkani-Tur, and Oliver Lemon. Let's DiSCoH: collecting an annotated open corpus with dialogue acts and reward signals for

natural language helpdesks. In *2006 IEEE Spoken Language Technology Workshop*, Palm Beach, Aruba, pages 218–221. IEEE, 2006.

[20] Magdalena Andrzejewska and Agnieszka Skawińska. Examining students' cognitive effort during program comprehension – an eye tracking approach. In *AIED 2020, Proceedings of the 21st International Conference on Artificial Intelligence in Education*, Ifrane, Morocco, July 2020.

[21] Florencia K. Anggoro, Nancy L. Stein, and Benjamin D. Jee. Cognitive factors that influence children's learning from a multimedia science lesson. *International Electronic Journal of Elementary Education*, 5(1):93–108, 2017.

[22] Yuichiro Anzai and Herbert A. Simon. The theory of learning by doing. *Psychological Review*, 86(2):124, 1979.

[23] Noah Arthurs and A.J. Alvero. Whose Truth is the "Ground Truth"? College Admissions Essays and Bias in Word Vector Evaluation Methods. In Anna N. Rafferty, Jacob Whitehill, Violetta Cavalli-Sforza, and Cristobal Romero, editors, *Proceedings of The 13th International Conference on Educational Data Mining (EDM 2020)*, Online, pages 342–349, 2020.

[24] Ron Artstein and Massimo Poesio. Inter-coder agreement for computational linguistics. *Computational Linguistics*, 34(4):555–596, 2008.

[25] Pino G. Audia and Edwin A. Locke. Benefiting from negative feedback. *Human Resource Management Review*, 13(4):631–646, 2003.

[26] John L. Austin. *How to do Things with Words*. Oxford University Press, Oxford, 1962.

[27] Robert W. Bailey. *The Effect of Stimulus Duration and Error-stage on the Ability to Self-detect Errors*. PhD thesis, Rice University, 1978.

[28] Michael Ball, Lauren Mock, Daniel D. Garcia, Tiffany Barnes, Marnie Hill, Alexandra Milliken, Joshua Paley, Efrain Lopez, and Jason Bohrer. The beauty and joy of computing curriculum and teacher professional development. In *Proceedings of the 51st ACM Technical Symposium on Computer Science Education*, Online, pages 1398–1398, 2020.

[29] Alexandre de A. Barbosa, Evandro de B. Costa, and Patrick H. Brito. Adaptive Clustering of Codes for Assessment in Introductory Programming Courses. In *International Conference on Intelligent Tutoring Systems*, pages 13–22. Springer, Montreal, Quebec, Canada, 2018.

[30] Lecia J. Barker, Charlie McDowell, and Kimberly Kalahar. Exploring Factors that Influence Computer Science Introductory Course Students to Persist in the Major. In *Proceedings of the 40th ACM Technical*

Symposium on Computer Science Education, SIGCSE '09, pages 153–157, New York, NY, USA, 2009. ACM.

[31] Tiffany Barnes, Kristy Elizabeth Boyer, Sharon I-Han Hsiao, Nguyen-Thinh Le, and Sergey Sosnovsky. Preface for the special issue on AI-supported education in Computer Science. *International Journal of Artificial Intelligence in Education*, 27(1):1–4, 2017.

[32] Tiffany Barnes and John C. Stamper. Toward the extraction of production rules for solving logic proofs. In *AIED 2007, Proceedings of the 13th International Conference on Artificial Intelligence in Education, Educational Data Mining Workshop*, pages 11–20, Marina Del Rey, CA, July 2007.

[33] Tiffany Barnes and John C. Stamper. Toward automatic hint generation for logic proof tutoring using historical student data. In *ITS 2008, The 9th International Conference on Intelligent Tutoring Systems*, pages 373–382, Montreal, Canada, June 2008.

[34] Devon Barrow, Antonija Mitrović, Stellan Ohlsson, and Michael Grimley. Assessing the impact of positive feedback in constraint-based tutors. In *ITS 2008, The 9th International Conference on Intelligent Tutoring Systems*, pages 250–259, Montreal, Canada, June 2008. Berlin, Germany, Springer Verlag.

[35] Theresa Beaubouef and John Mason. Why the high attrition rate for Computer Science students: some thoughts and observations. *SIGCSE Bull.*, 37(2):103–106, 2005.

[36] William Bechtel and George Graham, editors. *A Companion to Cognitive Science*. Blackwell Companions to Philosophy. Blackwell, 1998.

[37] Mordechai Ben-Ari. Constructivism in computer science education. *SIGCSE Bull.*, 30(1):257–261, 1998.

[38] Emily M. Bender and Alexander Koller. Climbing towards NLU: On meaning, form, and understanding in the age of data. In *Proceedings of the 58th Annual Meeting of the Association for Computational Linguistics*, pages 5185–5198, Online, Jul 2020. Association for Computational Linguistics.

[39] Sylvia Beyer. Why are women underrepresented in Computer Science? Gender differences in stereotypes, self-efficacy, values, and interests and predictors of future CS course-taking and grades. *Computer Science Education*, 24(2-3):153–192, 2014.

[40] Neeraj Bhatnagar and Jack Mostow. On-line learning from search failures. *Machine Learning*, 15(1):69–117, 1994.

[41] Benjamin S. Bloom. The 2 Sigma problem: The search for methods of group instruction as effective as one-to-one tutoring. *Educational Researcher*, 13:4–16, 1984.

[42] Nigel Bosch and Sidney D'Mello. The affective experience of novice computer programmers. *International Journal of Artificial Intelligence in Education*, 27(1):181–206, 2017.

[43] Randall W. Bower. *An investigation of a manipulative simulation in the learning of recursive programming.* PhD thesis, Iowa State University, 1998.

[44] Kristy Elizabeth Boyer, Robert Phillips, Eun Young Ha, Michael D. Wallis, Mladen A. Vouk, and James C. Lester. Modeling dialogue structure with adjacency pair analysis and Hidden Markov Models. In *NAACL 2009, Annual Conference of the North American Chapter of the Association for Computational Linguistics, Companion Volume: Short Papers*, pages 49–52. Association for Computational Linguistics, Boulder, CO, USA, 2009.

[45] Kristy Elizabeth Boyer, Robert Phillips, Amy Ingram, Eun Young Ha, Michael D. Wallis, Mladen Vouk, and James C. Lester. Characterizing the effectiveness of tutorial dialogue with Hidden Markov Models. In *Intelligent Tutoring Systems*, pages 55–64. Springer, 2010.

[46] Keith Brawner and Arthur Graesser. Natural language, discourse, and conversational dialogues within intelligent tutoring systems: A review. In Robert A. Sottilare, Arthur Graesser, Xiangen Hu, and Heather Holden, editors, *Design Recommendations for Intelligent Tutoring Systems*, volume 2: Instructional Management, pages 189–204. U.S. Army Research Laboratory, 2014.

[47] Julien Broisin, Rémi Venant, and Philippe Vidal. Lab4CE: a remote laboratory for computer education. *International Journal of Artificial Intelligence in Education*, 27(1):154–180, 2017.

[48] John Seely Brown and Richard R Burton. Diagnostic models for procedural bugs in basic mathematical skills. *Cognitive Science*, 2(2):155–192, 1978.

[49] Neil C. C. Brown and Amjad Altadmri. Novice Java Programming Mistakes: Large-Scale Data vs. Educator Beliefs. *ACM Trans. Comput. Educ.*, 17(2), May 2017.

[50] Jeff Bulmer, Angie Pinchbeck, and Bowen Hui. Visualizing code patterns in novice programmers. In *Proceedings of the 23rd Western Canadian Conference on Computing Education*, WCCCE '18, New York, NY, USA, 2018. Association for Computing Machinery.

[51] Harry Bunt, Jan Alexandersson, Jae-Woong Choe, Alex Chengyu Fang, Koiti Hasida, Volha Petukhova, Andrei Popescu-Belis, and David R. Traum. ISO 24617-2: A semantically-based standard for dialogue annotation. In *LREC 2012, Proceedings of the International Conference on Language Resources and Evaluation*, Istanbul, Turkey, pages 430–437, 2012.

[52] Harry Bunt, Volha Petukhova, David R. Traum, and Jan Alexandersson. Dialogue act annotation with the ISO 24617-2 standard. In *Multimodal interaction with W3C standards*, pages 109–135. Springer, 2017.

[53] Bruce D. Burns and Keith J. Holyoak. Competing models of analogy: ACME versus Copycat. In *Proceedings of the Sixteenth Annual Conference of the Cognitive Science Society*, pages 100–105. Routledge, 2019.

[54] Jill Burstein, Martin Chodorow, and Claudia Leacock. Criterion^{SM} Online Essay Evaluation: An Application for Automated Evaluation of *Proc. 15th Conference on Innovative Applications of Artificial Intelligence*, Acapulco, Mexico, In *IAAI*, pages 3–10, 2003.

[55] Whitney L. Cade, Jessica L. Copeland, Natalie K. Person, and Sidney K. D'Mello. Dialogue modes in expert tutoring. In *ITS 2008, the 9th International Conference on Intelligent Tutoring Systems*, pages 470–479, Montreal, Canada, June 2008. Springer.

[56] Jaime R. Carbonell. AI in CAI: An Artificial-Intelligence approach to computer-assisted instruction. *IEEE Transactions on Man-Machine Systems*, 11(4):190–202, 1970.

[57] Jean Carletta. Assessing agreement on classification tasks: the Kappa statistic. *Computational Linguistics*, 22(2):249–254, 1996.

[58] Jean Carletta, Amy Isard, Stephen Isard, Jacqueline C. Kowtko, Gwyneth Doherty-Sneddon, and Anne H. Anderson. The reliability of a dialogue structure coding scheme. *Computational Linguistics*, 23(1):13–31, 1997.

[59] Brian Carr and Ira P. Goldstein. Overlays: A theory of modelling for computer aided instruction. Technical report, Massachusetts Institute of Technology, Cambridge, Artificial Intelligence Laboratory, 1977.

[60] Jacqueline G. Cavazos, P. Jonathon Phillips, Carlos D. Castillo, and Alice J. O'Toole. Accuracy comparison across face recognition algorithms: Where are we on measuring race bias? *IEEE Transactions on Biometrics, Behavior, and Identity Science*, 3(1): 101–111, 2021.

[61] Carol K.K. Chan. Peer collaboration and discourse patterns in learning from incompatible information. *Instructional Science*, 29(6):443–479, 2001.

[62] Limin Chen, Zhiwen Tang, and Grace Hui Yang. Balancing reinforcement learning training experiences in interactive information retrieval. In *Proceedings of the 43rd International ACM Conference on Research and Development in Information Retrieval (SIGIR)*, pages 1525–1528, Xi'an, China, 2020.

[63] Lin Chen, Barbara Di Eugenio, Davide Fossati, Stellan Ohlsson, and David Cosejo. Exploring effective dialogue act sequences in one-on-one computer science tutoring dialogues. In *Proceedings of the Sixth Workshop on Innovative Use of NLP for Building Educational Applications*, pages 65–75, Portland, Oregon, June 2011. Association for Computational Linguistics.

[64] Lin Chen, Maria Javaid, Barbara Di Eugenio, and Miloš Žefran. The roles and recognition of haptic-ostensive actions in collaborative multimodal human-human dialogues. *Computer Speech & Language*, 32:201–231, Nov. 2015.

[65] Xingliang Chen, Antonija Mitrović, and Moffat Mathews. Learning from worked examples, erroneous examples and problem solving: Towards adaptive selection of learning activities. *IEEE Transactions on Learning Technologies*, 13(1):135–149, 2019.

[66] Michelene T. H. Chi. Active-constructive-interactive: A conceptual framework for differentiating learning activities. *Topics in Cognitive Science*, 1:73–105, 2009.

[67] Michelene T. H. Chi, Miriam Bassok, Matthew W. Lewis, Peter Reimann, and Robert Glaser. Self-explanations: How students study and use examples in learning to solve problems. *Cognitive Science*, 13(2):145–182, 1989.

[68] Michelene T. H. Chi, Nicholas de Leeuw, Mei-Hung Chiu, and Christian LaVancher. Eliciting self-explanations improves understanding. *Cognitive Science*, 18(3):439–477, 1994.

[69] Michelene T. H. Chi and Muhsin Menekse. Dialogue patterns that promote learning. In L. B. Resnick, C. Asterhan, and S. N. Clarke, editors, *Socializing intelligence through academic talk and dialogue*, Chapter 21, pages 263–274. AERA, Washington, DC, 2015.

[70] Michelene T.H. Chi. Quantifying qualitative analyses of verbal data: A practical guide. *Journal of the Learning Sciences*, 6(3):271–315, 1997.

[71] Michelene T.H. Chi and Marguerite Roy. How adaptive is an expert human tutor? In *ITS 2010, Proceedings of the 10th International Conference on Intelligent Tutoring Systems*, pages 401–412. Springer, 2010.

[72] Anna L. Choi, Sylvaine Cordier, Pál Weihe, and Philippe Grand-jean. Negative confounding in the evaluation of toxicity: The case of methylmercury in fish and seafood. *Critical Reviews in Toxicology*, 38(10):877–893, 2008.

[73] Konstantina Chrysafiadi and Maria Virvou. Student modeling approaches: A literature review for the last decade. *Expert Systems with Applications*, 40(11):4715–4729, 2013.

[74] Jennifer Chu-Carroll and Michael K. Brown. An evidential model for tracking initiative in collaborative dialogue interactions. *User Modeling and User-Adapted Interaction*, 8(3–4):215–253, September 1998.

[75] William J Clancey. From GUIDON to NEOMYCIN and HERACLES in twenty short lessons. *AI Magazine*, 7(3):40–40, 1986.

[76] Ruth C. Clark, Frank Nguyen, and John Sweller. *Efficiency in Learning: Evidence-Based Guidelines to Manage Cognitive Load*. John Wiley & Sons, 2011.

[77] Jacob Cohen. A coefficient of agreement for nominal scales. *Educational and Psychological Measurement*, 20:37–46, 1960.

[78] Philip Cohen and Hector Levesque. Persistence, Intention and Commitment. In P. Cohen, J. Morgan, and M. Pollack, editors, *Intentions in Communication*. MIT Press, 1990.

[79] Philip Cohen and Hector Levesque. Rational Interaction as the Basis for Communication. In P. Cohen, J. Morgan, and M. Pollack, editors, *Intentions in Communication*. MIT Press, 1990.

[80] Cristina Conati, Abigail Gertner, and Kurt VanLehn. Using bayesian networks to manage uncertainty in student modeling. *Journal of User Modeling and User-Adapted Interaction*, 12(4):371–417, 2002. Winner of the 2002 James Chen Annual Award for Best UMUAI Paper.

[81] Cristina Conati and Kurt VanLehn. Teaching meta-cognitive skills: implementation and evaluation of a tutoring system to guide self-explanation while learning from examples. In *AIED 1999, Proceedings of the 9th International Conference of Artificial Intelligence and Education*, Le Mans, France, 1999.

[82] Stephen Cooper, Wanda Dann, and Randy Pausch. Alice: a 3-D tool for introductory programming concepts. *Journal of Computing Sciences in Colleges*, 15(5):107–116, 2000.

[83] Albert T. Corbett and John R. Anderson. The effect of feedback control on learning to program with the Lisp tutor. In *Proceedings of the Twelfth Annual Conference of the Cognitive Science Society*, pages 796–803, Cambridge, MA, 1990.

[84] Albert T. Corbett, John R. Anderson, Arthur C. Graesser, Kenneth R. Koedinger, and Kurt VanLehn. Third generation computer tutors: Learn from or ignore human tutors? In *Proceedings of the 1999 Conference of Computer-Human Interaction*, pages 85–86. ACM Press, New York, 1999.

[85] Mark G. Core, Johanna D. Moore, and Claus Zinn. The role of initiative in tutorial dialogue. In *EACL '03: Proceedings of the Tenth Conference of the European Chapter of the Association for Computational Linguistics*, pages 67–74, Morristown, NJ, USA, 2003. Association for Computational Linguistics.

[86] Thomas H. Cormen, Charles E. Leiserson, Ronald L. Rivest, and Clifford Stein. *Introduction to Algorithms, 2nd edition*. MIT Press, McGraw-Hill Book Company, New York, NY, 2000.

[87] Andrew Corrigan-Halpern and Stellan Ohlsson. Failure to learn from negative feedback in a hierarchical adaptive system. In *The Fourth International Conference on Computer Modeling*, Fairfax, VA, USA, 2001.

[88] Andrew Corrigan-Halpern and Stellan Ohlsson. Feedback effects in the acquisition of a hierarchical skill. In *Proceedings of the 24th Annual Conference of the Cognitive Science Society*, Fairfax, VA, USA, 2002.

[89] Corinna Cortes and Vladimir Vapnik. Support-vector networks. *Machine Learning*, 20(3):273–297, 1995.

[90] Keeley Crockett, Annabel Latham, and Nicola Whitton. On predicting learning styles in conversational Intelligent Tutoring Systems using fuzzy decision trees. *International Journal of Human-Computer Studies*, 97:98–115, 2017.

[91] Tyne Crow, Andrew Luxton-Reilly, and Burkhard Wuensche. Intelligent Tutoring Systems for programming education: a systematic review. In *Proceedings of the 20th Australasian Computing Education Conference*, Brisbane, Queensland, Australia, pages 53–62, 2018.

[92] Jason J. Dahling and Christopher L. Ruppel. Learning goal orientation buffers the effects of negative normative feedback on test self-efficacy and reattempt interest. *Learning and Individual Differences*, 50:296–301, 2016.

[93] Crina I. Damşa. The multi-layered nature of small-group learning: Productive interactions in object-oriented collaboration. *International Journal of Computer-Supported Collaborative Learning*, pages 1–35, 2014.

[94] Garrett Dancik and Amruth N. Kumar. A tutor for counter-controlled loop concepts and its evaluation. In *Proceedings of the 33rd ASEE/IEEE Frontiers in Education Conference*, Westminster, CO, USA, pages 7–9, 2003.

[95] Wanda Dann, Stephen Cooper, and Randy Pausch. Using visualization to teach novices recursion. In *6th Conference on Innovation and Technology in Computer Science Education*, ITiCSE '01, pages 109–112, New York, NY, USA, 2001. ACM.

[96] Tim DeClue, Jeff Kimball, Baochuan Lu, and James Cain. Five focused strategies for increasing retention in Computer Science 1. *Journal of Computing Sciences in Colleges*, 26(5):252–258, 2011.

[97] Scott Deerwester, Susan T. Dumais, George W. Furnas, Thomas K. Landauer, and Richard Harshman. Indexing by Latent Semantic Analysis. *Journal of the American Society for Information Science*, 41:391–407, 1990.

[98] Peter F. Delaney, Lynne M. Reder, James J. Staszewski, and Frank E. Ritter. The strategy-specific nature of improvement: The power law applies by strategy within task. *Psychological Science*, 9(1):1–7, 1998.

[99] Peter J. Denning. Beyond computational thinking. *Communications of the ACM*, 52(6):28–30, June 2009.

[100] Jacob Devlin, Ming-Wei Chang, Kenton Lee, and Kristina Toutanova. BERT: Pre-training of deep bidirectional transformers for language understanding. In *Proceedings of the 2019 Conference of the North American Chapter of the Association for Computational Linguistics: Human Language Technologies, Volume 1 (Long and Short Papers)*, pages 4171–4186, Minneapolis, Minnesota, June 2019. Association for Computational Linguistics.

[101] John Dewey. *Democracy and Education: An Introduction to the Philosophy of Education.* (Originally published in 1916). New York: WLC Books, 2009.

[102] Barbara Di Eugenio, Davide Fossati, Stellan Ohlsson, and David Cosejo. Towards explaining effective tutorial dialogues. In *CogSci 2009, the Annual Meeting of the Cognitive Science Society*, Amsterdam, the Netherlands, July 2009.

[103] Barbara Di Eugenio and Michael Glass. The Kappa statistic: A second look. *Computational Linguistics*, 30(1):95–101, 2004. Squib.

[104] Barbara Di Eugenio, Xin Lu, Trina C. Kershaw, Andrew Corrigan-Halpern, and Stellan Ohlsson. Positive and negative verbal feedback for

intelligent tutoring systems. In *AIED 2005, the 12th International Conference on Artificial Intelligence in Education*, Amsterdam, The Netherlands, 2005. Poster.

[105] Sidney D'Mello and Arthur C. Graesser. AutoTutor and affective AutoTutor: Learning by talking with cognitively and emotionally intelligent computers that talk back. *ACM Transactions on Interactive Intelligent Systems*, 2(4):1–39, 2013.

[106] Jie Du, Hayden Wimmer, and Roy Rada. Hour of code: A case study. *Information Systems Education Journal*, 16(1):51–59, 2018.

[107] Pat Dugard and John Todman. Analysis of Pre-test-Post-test Control Group Designs in Educational Research. *Educational Psychology*, 15(2):181–198, 1995.

[108] Myroslava Dzikovska, Natalie Steinhauser, Elaine Farrow, Johanna Moore, and Gwendolyn Campbell. BEETLE II: Deep natural language understanding and automatic feedback generation for intelligent tutoring in basic electricity and electronics. *International Journal of Artificial Intelligence in Education*, 24(3):284–332, 2014.

[109] Jeffrey Edgington. Teaching and viewing recursion as delegation. *Journal of Computing Sciences in Colleges*, 23(1):241–246, 2007.

[110] Mark Elsom-Cook. Student modelling in Intelligent Tutoring Systems. *Artificial Intelligence Review*, 7(3-4):227–240, 1993.

[111] Martha W. Evens and Joel A. Michael. *One-on-one Tutoring by Humans and Machines*. Mahwah, NJ: Lawrence Erlbaum Associates, 2006.

[112] Martha W. Evens, John Spitkovsky, Patrick Boyle, Joel A. Michael, and Allen A. Rovick. Synthesizing Tutorial Dialogues. In *Proceedings of the Fifteenth Annual Conference of the Cognitive Science Society*, pages 137–140, Hillsdale, New Jersey, 1993. Lawrence Erlbaum Associates.

[113] Aysu Ezen-Can and Kristy Elizabeth Boyer. In-Context Evaluation of Unsupervised Dialogue Act Models for Tutorial Dialogue. In *Proceedings of the SIGDIAL 2013 Conference*, pages 324–328, Metz, France, August 2013. Association for Computational Linguistics.

[114] Aysu Ezen-Can and Kristy Elizabeth Boyer. A tutorial dialogue system for real-time evaluation of unsupervised dialogue act classifiers: Exploring system outcomes. In *International Conference on Artificial Intelligence in Education*, pages 105–114. Springer, Madrid, Spain, 2015.

[115] Aysu Ezen-Can and Kristy Elizabeth Boyer. Understanding student language: An unsupervised dialogue act classification approach. *Journal of Educational Data Mining (JEDM)*, 7(1):51–78, 2015.

[116] Brian Falkenhainer, Kenneth D. Forbus, and Dedre Gentner. The structure-mapping engine: Algorithm and examples. *Artificial Intelligence*, 41(1):1–63, 1989.

[117] Nickolas J.G. Falkner and Katrina E. Falkner. A fast measure for identifying at-risk students in Computer Science. In *Proceedings of the Ninth Annual International Conference on International Computing Education Research*, Auckland, New Zealand, pages 55–62. ACM, 2012.

[118] Emmanuel Ferreira and Fabrice Lefevre. Reinforcement-learning based dialogue system for human–robot interactions with socially-inspired rewards. *Computer Speech & Language*, 34(1):256–274, 2015.

[119] Charles B. Ferster and Burrhus Frederic Skinner. *Schedules of Reinforcement*. Appleton-Century-Crofts, 1957.

[120] Allan Fisher, Jane Margolis, and Faye Miller. Undergraduate women in Computer Science: Experience, motivation and culture. *ACM SIGCSE Bulletin*, 29(1):106–110, 1997.

[121] Peter W. Foltz, Darrell Laham, and Thomas K. Landauer. The intelligent essay assessor: Applications to educational technology. *Interactive Multimedia Electronic Journal of Computer-Enhanced Learning*, 1(2), 1999.

[122] Kate Forbes-Riley and Diane J. Litman. Modelling user satisfaction and student learning in a spoken dialogue tutoring system with generic, tutoring, and user affect parameters. In *Proceedings of the Human Language Technology Conference of the North American Chapter of the Association of Computational Linguistics*, New York, NY, USA, pages 264–271. Association for Computational Linguistics, 2006.

[123] Kate Forbes-Riley and Diane J. Litman. Analyzing dependencies between student certainness states and tutor responses in a spoken dialogue corpus. In Laila Dybkjær and Wolfgang Minker, editors, *Recent Trends in Discourse and Dialogue*, number 39 in Text, Speech and Language Technology, pages 275–304. Springer Netherlands, 2008.

[124] Kate Forbes-Riley and Diane J. Litman. When does disengagement correlate with performance in spoken dialog computer tutoring? *International Journal of Artificial Intelligence in Education*, 22(1–2):39–58, 2013.

[125] Kenneth D. Forbus, Ronald W. Ferguson, Andrew Lovett, and Dedre Gentner. Extending SME to handle large-scale cognitive modeling. *Cognitive Science*, 41(5):1152–1201, 2017.

[126] Kenneth D. Forbus, Dedre Gentner, and Keith Law. MAC/FAC: A model of similarity-based retrieval. *Cognitive Science*, 19(2):141–205, 1995.

[127] Gary Ford. An implementation-independent approach to teaching recursion. *SIGCSE Bull.*, 16(1):213–216, January 1984.

[128] Davide Fossati. *Automatic Modeling of Procedural Knowledge and Feedback Generation in a Computer Science Tutoring System*. PhD thesis, University of Illinois - Chicago, Summer 2009.

[129] Davide Fossati, Barbara Di Eugenio, Christopher Brown, and Stellan Ohlsson. Learning Linked Lists: Experiments with the iList System. In *ITS 2008, the 9th International Conference on Intelligent Tutoring Systems*, Montreal, Canada, 2008.

[130] Davide Fossati, Barbara Di Eugenio, Christopher Brown, Stellan Ohlsson, David Cosejo, and Lin Chen. Supporting Computer Science curriculum: Exploring and learning linked lists with iList. *IEEE Transactions on Learning Technologies, Special Issue on Real-World Applications of Intelligent Tutoring Systems*, 2(2):107–120, April-June 2009.

[131] Davide Fossati, Barbara Di Eugenio, Stellan Ohlsson, Christopher Brown, and Lin Chen. Data driven automatic feedback generation in the iList intelligent tutoring system. *Technology, Instruction, Cognition and Learning (TICL), Special Issue on the Role of Data in Instructional Processes*, 10:5–26, 2015.

[132] Davide Fossati, Barbara Di Eugenio, Stellan Ohlsson, Christopher Brown, Lin Chen, and David Cosejo. I learn from you, you learn from me: How to make iList learn from students. In *AIED09, the 14th International Conference on Artificial Intelligence in Education*, Brighton, Great Britain, 2009.

[133] Barbara A. Fox. *The Human Tutorial Dialogue Project: Issues in the design of instructional systems*. Lawrence Erlbaum Associates, Hillsdale, NJ, 1993.

[134] Reva K. Freedman. Plan-based dialogue management in a physics tutor. In *Proceedings of the Sixth Applied Natural Language Conference*, Seattle, WA, May 2000.

[135] Yoav Freund and Robert E. Schapire. Large margin classification using the perceptron algorithm. *Machine Learning*, 37(3):277–296, 1999.

[136] Ursula Fuller, Colin G. Johnson, Tuukka Ahoniemi, Diana Cukierman, Isidoro Hernán-Losada, Jana Jackova, Essi Lahtinen, Tracy L. Lewis, Donna McGee Thompson, Charles Riedesel, and Errol Thompson. Developing a Computer Science-specific Learning Taxonomy. *SIGCSE Bull.*, 39(4):152–170, 2007.

[137] Crystal Furman, Owen Astrachan, Daniel D. Garcia, David Musicant, and Jennifer Rosato. CS Principles Higher Education Pathways. In *Proceedings of the 50th ACM Technical Symposium on Computer Science Education*, pages 498–499, 2019.

[138] Soniya Gadgil and Timothy J. Nokes. Analogical scaffolding in collaborative learning. In *Proceedings of the Annual Conference of the Cognitive Science Society*, pages 3115–3120, 2010.

[139] Robert M. Gagné. *The conditions of learning.* Holt, Rinehart and Winston, New York, NY, USA, 1965.

[140] Judith Gal-Ezer and David Harel. What (else) should CS educators know? *Communications of the ACM*, 41(9):77–84, 1998.

[141] Charles R. Gallistel. *The Organization of Action: A new synthesis.* Psychology Press, 1980.

[142] Daniel D. Garcia, Jennifer Campbell, Rebecca Dovi, and Cay Horstmann. Rediscovering the passion, beauty, joy, and awe: making computing fun again, part 7. In *Proceedings of the 45th ACM Technical Symposium on Computer Science Education*, SIGCSE '14, Atlanta, GA, USA, pages 273–274. ACM, 2014.

[143] Daniel D. Garcia, Moses Charikar, Eboney Hearn, Ed Lazowska, and Jonathan Reynolds. Institutions share successes, failures, and advice in moving the diversity needle. In *Proceedings of the 51st ACM Technical Symposium on Computer Science Education*, Online, SIGCSE '20 pages 331–332, 2020.

[144] Daniel D. Garcia, Robb Cutler, Zachary Dodds, Eric Roberts, and Alison Young. Rediscovering the passion, beauty, joy, and awe: making computing fun again, continued. In *Proceedings of the 40th ACM Technical Symposium on Computer Science Education*, SIGCSE '09, Chattanooga, TN, USA, pages 65–66. ACM, 2009.

[145] Daniel D. Garcia, Brian Harvey, and Tiffany Barnes. The beauty and joy of computing. *ACM Inroads*, 6(4):71–79, 2015.

[146] Raul Vicente Garcia, Lukasz Wandzik, Louisa Grabner, and Joerg Krueger. The harms of demographic bias in deep face recognition research. In *International Conference on Biometrics (ICB)*, Crete, Greece, pages 1–6. IEEE, 2019.

[147] Alison Garton. *Social Interaction and the Development of Language and Cognition.* Psychology Press, 1995.

[148] Albert Gatt and Emiel Krahmer. Survey of the state of the art in Natural Language Generation: Core tasks, applications and evaluation. *Journal of Artificial Intelligence Research*, 61:65–170, 2018.

[149] Dedre Gentner. Structure-mapping: A theoretical framework for analogy. *Cognitive Science*, 7:155–170, 1983.

[150] Dedre Gentner. Analogy. In William Bechtel and George Graham, editors, *A Companion to Cognitive Science*, pages 107–113. Wiley Online Library, 2017.

[151] Dedre Gentner and Julie Colhoun. Analogical processes in human thinking and learning. In Britt Glatzeder, Vinod Goel, and Albrecht von Müller, editors, *Towards a Theory of Thinking: Building Blocks for a Conceptual Framework*, pages 35–48. Springer-Verlag, Berlin Heidelberg, 2010.

[152] Dedre Gentner and Christian Hoyos. Analogy and abstraction. *Topics in Cognitive Science*, 9(3):672–693, 2017.

[153] Dedre Gentner and Arthur B. Markman. Analogy–Watershed or Waterloo? Structural alignment and the development of connectionist models of analogy. In *Advances in Neural Information Processing Systems 5*, pages 855–862, 1993.

[154] Dedre Gentner and Linsey Smith. Analogical reasoning. In V.S. Ramachandran, editor, *Encyclopedia of Human Behavior*, volume 1, pages 130–136. Academic Press, 2012.

[155] Alex Gerdes, Bastiaan Heeren, Johan Jeuring, and L. Thomas van Binsbergen. Ask-Elle: an adaptable programming tutor for Haskell giving automated feedback. *International Journal of Artificial Intelligence in Education*, 27(1):65–100, 2017.

[156] Mary L. Gick and Keith J. Holyoak. Analogical problem solving. *Cognitive Psychology*, 12(3):306–55, 1980.

[157] Mary L. Gick and Keith J. Holyoak. Schema induction and analogical transfer. *Cognitive Psychology*, 15(1):1–38, 1983.

[158] David Ginat and Eyal Shifroni. Teaching recursion in a procedural environment - How much should we emphasize the computing model? *SIGCSE Bull.*, 31(1):127–131, March 1999.

[159] Michael Glass. Some phenomena handled by the CIRCSIM Tutor version 3 input understander. In *FLAIRS97, Proceedings of the 9th International Florida Artificial Intelligence Research Symposium*, Daytona Beach, Florida, USA, 1997.

[160] Joanna Goode. If You Build Teachers, Will Students Come? The Role of Teachers in Broadening Computer Science Learning for Urban Youth. *Journal of Educational Computing Research*, 36(1):65–88, 2007.

[161] Joanna Goode. Connecting K-16 Curriculum & Policy: Making Computer Science Engaging, Accessible, and Hospitable for Underrepresented Students. In *Proceedings of the 41st ACM Technical Symposium on Computer Science Education*, SIGCSE '10, pages 22–26, New York, NY, USA, 2010. ACM.

[162] Joanna Goode and Jane Margolis. What is Computer Science, Anyway?: Deepening Urban Teachers' Understandings of Computer Science and Working Towards an Engaging Pedagogy . In Richard Ferdig, Caroline Crawford, Roger Carlsen, Niki Davis, Jerry Price, Roberta Weber, and Dee Anna Willis, editors, *Proceedings of the Society for Information Technology & Teacher Education International Conference*, pages 814–819, Atlanta, GA, USA, 2004. Association for the Advancement of Computing in Education (AACE).

[163] Joanna Goode, Max Skorodinsky, Jill Hubbard, and James Hook. Computer Science for Equity: Teacher Education, Agency, and Statewide Reform. In *Frontiers in Education*, 2020.

[164] Tina Götschi, Ian Sanders, and Vashti Galpin. Mental models of recursion. *SIGCSE Bulletin*, 35(1):346–350, January 2003.

[165] Arthur C. Graesser. Dialogue Patterns and Feedback Mechanisms during Naturalistic Tutoring. In *Proceedings of the Cognitive Science Society Annual Meeting*, Boulder, CO, USA, pages 126–130, 1993.

[166] Arthur C. Graesser, Shulan Lu, G. Tanner Jackson, Heather H. Mitchell, Mathew Ventura, Andrew Olney, and Max M. Louwerse. AutoTutor: A tutor with dialogue in natural language. *Behavioral Research Methods, Instruments, and Computers*, 36:180–193, 2004.

[167] Arthur C. Graesser and Natalie K. Person. Question asking during tutoring. *American Educational Research Journal*, 31(1):104–137, 1994.

[168] Arthur C. Graesser, Natalie K. Person, Derek Harter, and The Tutoring Research Group. Teaching Tactics and Dialog in AutoTutor. *International Journal of Artificial Intelligence in Education*, 12:257–279, 2001.

[169] Arthur C. Graesser, Natalie K. Person, Zhijun Lu, Moon Gee Jeon, and Bethany McDaniel. Learning while holding a conversation with a computer. In L. Pytlik Zillig, M. Bodvarsson, and R. Brunin, editors, *Technology-based education: Bringing Researchers and Practitioners Together*. Information Age Publishing, 2005.

[170] Arthur C. Graesser, Natalie K. Person, and Joseph P. Magliano. Collaborative dialogue patterns in naturalistic one-to-one tutoring. *Applied Cognitive Psychology*, 9:495–522, 1995.

[171] Arthur C. Graesser, Kurt VanLehn, Carolyn Penstein Rosé, Pamela W. Jordan, and Derek Harter. Intelligent tutoring systems with conversational dialogue. *AI Magazine*, 22(4):39–52, 2001.

[172] Arthur C. Graesser, Katja Wiemer-Hastings, Peter Wiemer-Hastings, Roger Kreuz, and the Tutoring Research Group. AutoTutor: A simulation of a human tutor. *Journal of Cognitive Systems Research*, 1(1), 1999.

[173] Joseph Grafsgaard, Joseph B. Wiggins, Kristy Elizabeth Boyer, Eric N. Wiebe, and James Lester. Automatically recognizing facial expression: Predicting engagement and frustration. In *Educational Data Mining 2013*, 2013.

[174] Nicholas Green. *Example Based Pedagogical Strategies in a Computer Science Intelligent Tutoring System*. PhD thesis, University of Illinois at Chicago, 2017.

[175] Nick Green, Barbara Di Eugenio, Rachel Harsley, Davide Fossati, Omar AlZoubi, and Mehrdad Alizadeh. Student Behavior with Worked-out Examples in a Computer Science Intelligent Tutoring System. In *International Conference on Educational Technologies*, Florianopolis, Santa Catarina, Brazil, November 2015.

[176] Foteini Grivokostopoulou, Isidoros Perikos, and Ioannis Hatzilygeroudis. An educational system for learning search algorithms and automatically assessing student performance. *International Journal of Artificial Intelligence in Education*, 27(1):207–240, 2017.

[177] Curry I. Guinn. An analysis of initiative selection in collaborative task-oriented discourse. *User Modeling and User-Adapted Interaction*, 8(3-4):255–314, 1998.

[178] Isabelle Guyon and André Elisseeff. An introduction to variable and feature selection. *Journal of Machine Learning Research*, 3(Mar):1157–1182, 2003.

[179] Richard R. Hake. Interactive-engagement versus traditional methods: A six-thousand-student survey of mechanics test data for introductory physics courses. *American journal of Physics*, 66(1):64–74, 1998.

[180] Mark Hall, Eibe Frank, Geoffrey Holmes, Bernhard Pfahringer, Peter Reutemann, and Ian H. Witten. The WEKA data mining software: An update. *SIGKDD Explorations*, 11(1), 2009.

[181] Sally Hamouda, Stephen H. Edwards, Hicham G. Elmongui, Jeremy V. Ernst, and Clifford A. Shaffer. RecurTutor: An interactive tutorial for learning recursion. *ACM Transactions on Computing Education (TOCE)*, 19(1):1, 2018.

[182] Rachel Harsley, Barbara Di Eugenio, Nicholas Green, Davide Fossati, and Sabita Acharya. Integrating support for collaboration in a computer science intelligent tutoring system. In Alessandro Micarelli, John Stamper, and Kitty Panourgia, editors, *ITS 2016, Proceedings of the 13th International Conference on Intelligent Tutoring Systems*, pages 227–233. Springer International Publishing, Zagreb, Croatia, June 2016.

[183] Rachel Harsley, Barbara Di Eugenio, Nick Green, and Davide Fossati. Enhancing an intelligent tutoring system to support student collaboration: Effects on learning and behavior. In *AIED 2017, Proceedings of the 18th International Conference on Artificial Intelligence in Education*, pages 519–522, Wuhan, China, 2017. Springer.

[184] Rachel Harsley, Davide Fossati, Barbara Di Eugenio, and Nicholas Green. Interactions of Individual and Pair Programmers with an Intelligent Tutoring System for Computer Science. In *SIGCSE 2017, Proceedings of the 48th ACM Technical Symposium on Computer Science Education*. ACM, Seattle, Washington, USA, March 2017.

[185] Rachel Harsley, Nick Green, Barbara Di Eugenio, Satabdi Aditya, Davide Fossati, and Omar Al Zoubi. Collab-ChiQat: A Collaborative Remaking of a Computer Science Intelligent Tutoring System. In *Proceedings of the 19th ACM Conference on Computer Supported Cooperative Work and Social Computing*, pages 281–284, New York, NY, USA, 2016. ACM.

[186] James Hartley. The effect of pre-testing on post-test performance. *Instructional Science*, 2(2):193–214, 1973.

[187] Peter Hastings, Simon Hughes, and M. Anne Britt. Active learning for improving machine learning of student explanatory essays. In *International Conference on Artificial Intelligence in Education*, pages 140–153. Springer, London. United Kingdom, 2018.

[188] Alexander G. Hauptmann and Alexander I. Rudnicky. Talking to computers: an empirical investigation. *International Journal of Man-Machine Studies*, 28(6):583–604, 1988.

[189] Robert G.M. Hausmann, Michelene T.H. Chi, and Marguerite Roy. Learning from collaborative problem solving: An analysis of three hypothesized mechanisms. In K.D Forbus, D. Gentner, and T. Regier, editors, *Proceedings of the 26th Annual Conference of the Cognitive Science Society*, Chicago, IL, USA, pages 547–552, Mahwah, NJ, 2004.

[190] Matthew J. Hays, H. Chad Lane, Daniel Auerbach, Mark G. Core, Dave Gomboc, and Milton Rosenberg. Feedback specificity and the learning of intercultural communication skills. In *AIED 2009, Proceedings of the 14th International Conference on Artificial Intelligence in Education*, Brighton, United Kingdom, pages 391–398, 2009.

[191] Andrew Heathcote, Scott Brown, and D.J.K. Mewhort. The power law repealed: The case for an exponential law of practice. *Psychonomic Bulletin & Review*, 7(2):185–207, 2000.

[192] Peter A. Heeman, Fan Yang, and Susan E. Strayer. Control in task-oriented dialogues. In *EUROSPEECH-2003*, pages 209–212, 2003.

[193] Neil T. Heffernan. Web-based evaluations showing both cognitive and motivational benefits of the Ms. Lindquist tutor. In *AIED03, Eleventh International Conference on Artificial Intelligence in Education*, Sydney, Australia, 2003.

[194] Neil T. Heffernan and Kenneth R. Koedinger. An intelligent tutoring system incorporating a model of an experienced human tutor. In *ITS'02, Proceedings of the 6th International Conference on Intelligent Tutoring Systems*, Biarritz, France and San Sebastian, Spain, 2002.

[195] Nicole Herbert, David Herbert, Erik Wapstra, Kristy de Salas, and Tina Acuña. An Exploratory Study of Factors Affecting Attrition within an ICT Degree. In *Proceedings of the Twenty-Second Australasian Computing Education Conference*, ACE'20, page 76–85, New York, NY, USA, 2020. Association for Computing Machinery.

[196] Stephanie Herppich, Jörg Wittwer, Matthias Nückles, and Alexander Renkl. Expertise amiss: Interactivity fosters learning but expert tutors are less interactive than novice tutors. *Instructional Science*, 44(3):205–219, 2016.

[197] Tatjana S. Hilbert, Silke Schworm, and A. Renkl. Learning from worked-out examples: The transition from instructional explanations to self-explanation prompts. *Instructional design for effective and enjoyable computer-supported learning*, pages 184–192, 2004.

[198] Douglas R. Hofstadter. Analogy as the core of cognition. In Dedre Gentner, Keith J. Holyoak, and Boicho N. Kokinov, editors, *The analogical mind: Perspectives from cognitive science*, pages 499–538. The MIT Press/Bradford Books, Cambridge, MA, 2001.

[199] Keith J. Holyoak and Paul Thagard. Analogical mapping by constraint satisfaction. *Cognitive Science*, 13(3):295–355, 1989.

[200] Keith J. Holyoak and Paul Thagard. *Mental leaps: Analogy in creative thought*. MIT Press, 1995.

[201] Donald L. Horowitz. Electoral systems: A primer for decision makers. *Journal of Democracy*, 14(4):115–127, 2003.

[202] Cynthia Howard (Kersey), Pamela W. Jordan, Barbara Di Eugenio, and Sandra Katz. Exploring initiative as a signal of knowledge co-construction during collaborative problem solving. *Cognitive Science*, 41:1422–1449, 2017.

[203] Cynthia Howard (Kersey), Pamela W. Jordan, Barbara Di Eugenio, and Sandra Katz. Shifting the load: A peer dialogue agent that encourages its human collaborator to contribute more to problem solving. *International Journal of Artificial Intelligence in Education*, 27(1):101–129, 2017.

[204] Iris Howley, Takayuki Kanda, Kotaro Hayashi, and Carolyn Penstein Rosé. Effects of social presence and social role on help-seeking and learning. In *Proceedings of the 2014 ACM/IEEE International Conference on Human-Robot Interaction*, pages 415–422. ACM, 2014.

[205] Iris K. Howley and Carolyn Penstein Rosé. Towards careful practices for automated linguistic analysis of group learning. *Journal of Learning Analytics*, 3(3):239–262, 2016.

[206] Gregory D. Hume, Joel A. Michael, Allen A. Rovick, and Martha W. Evens. Hinting as a tactic in one-on-one tutoring. *Journal of the Learning Sciences*, 5(1):23–47, 1996.

[207] John E. Hummel and Keith J. Holyoak. A symbolic-connectionist theory of relational inference and generalization. *Psychological Review*, 110(2):220–264, 2003.

[208] Christopher D. Hundhausen, Sarah A. Douglas, and John T. Stasko. A meta-study of algorithm visualization effectiveness. *Journal of Visual Languages and Computing*, 13(3):259–290, 2002.

[209] Natasha Jaques, Cristina Conati, Jason M. Harley, and Roger Azevedo. Predicting affect from gaze data during interaction with an Intelligent Tutoring System. In *ITS'14, Proceedings of the 12th International Conference on Intelligent Tutoring Systems*, pages 29–38. Springer, 2014.

[210] Finn V. Jensen. *An introduction to Bayesian networks*. Springer-Verlag, Berlin, Heidelberg, 1st edition, 1996.

[211] Youxuan Jiang, Jonathan K. Kummerfeld, and Walter S. Lasecki. Understanding task design trade-offs in crowdsourced paraphrase collection. In *Proceedings of the 55th Annual Meeting of the Association for Computational Linguistics (Volume 2: Short Papers)*, pages 103–109, Vancouver, Canada, July 2017. Association for Computational Linguistics.

[212] Randolph M. Jones and Pat Langley. A constrained architecture for learning and problem solving. *Computational Intelligence*, 21(4):480–502, 2005.

[213] Randolph M. Jones and Kurt VanLehn. Acquisition of children's addition strategies: A model of impasse-free, knowledge-level learning. *Machine Learning*, 16(1-2):11–36, 1994.

[214] Pamela W. Jordan, Patricia Albacete, and Sandra Katz. A comparison of tutoring strategies for recovering from a failed attempt during faded support. In *AIED 2018, Proceedings of the 19th International Conference on Artificial Intelligence in Education*, pages 212–224. Springer, 2018.

[215] Pamela W. Jordan and Barbara Di Eugenio. Control and initiative in collaborative problem solving dialogues. In *Working Notes of the AAAI Spring Symposium on Computational Models for Mixed Initiative*, pages 81–84, Menlo Park, CA, 1997.

[216] Pamela W. Jordan, Carolyn Penstein Rosé, and Kurt VanLehn. Tools for authoring tutorial dialogue knowledge. In Johanna D. Moore, Carol Luckhardt Redfield, and W. Lewis Johnson, editors, *AIED 2001, Proceedings of the 10th International Conference on Artificial Intelligence in Education*, pages 222–233, San Antonio, TX, May 2001. IOS Press.

[217] Daniel Jurafsky and James H. Martin. *Speech and Language Processing*. Pearson Education – Prentice Hall, third edition, 2021. In Progress. Available from https://web.stanford.edu/~jurafsky/slp3/.

[218] Hank Kahney. What do novice programmers know about recursion. In *Proceedings of the SIGCHI conference on Human Factors in Computing Systems*, pages 235–239. ACM, 1983.

[219] Slava Kalyuga, Paul Ayres, Paul Chandler, and John Sweller. The expertise reversal effect. *Educational Psychologist*, 38(1):23–31, 2003.

[220] Andreas Kappes, Gabriele Oettingen, and Hyeonju Pak. Mental contrasting and the self-regulation of responding to negative feedback. *Personality and Social Psychology Bulletin*, 38(7):845–857, 2012.

[221] Kuba Karpierz and Steven A Wolfman. Misconceptions and concept inventory questions for binary search trees and hash tables. In *Proceedings of the 45th ACM Technical Symposium on Computer Science Education*, Atlanta, GA, USA, pages 109–114, 2014.

[222] Sandra Katz. Gendered attrition at the undergraduate level. In Eileen Trauth, editor, *Encyclopedia of Gender and Information Technology*, pages 714–720. Idea Group Publishing, Hershey, Pennsylvania, 2006.

[223] Sandra Katz, Pamela W. Jordan, and Patricia Albacete. Exploring how to adaptively apply tutorial dialogue tactics. In *2016 IEEE 16th International Conference on Advanced Learning Technologies (ICALT)*, Austin, TX, USA, pages 36–38. IEEE, 2016.

[224] Judy Kay. Lifelong learner modeling for lifelong personalized pervasive learning. *IEEE Transactions on Learning Technologies*, 1(4):215–228, 2008.

[225] Colleen Kehoe, John Stasko, and Ashley Taylor. Rethinking the evaluation of algorithm animations as learning aids: an observational study. *International Journal of Human-Computer Studies*, 54(2):265–284, 2001.

[226] Cynthia Kersey, Barbara Di Eugenio, Pamela W. Jordan, and Sandra Katz. Modeling knowledge co-construction for peer learning interactions. In *ITS 2008, The 9th International Conference on Intelligent Tutoring Systems, Student Research Workshop*, Montreal, Canada, June 2008.

[227] Cynthia Kersey, Barbara Di Eugenio, Pamela W. Jordan, and Sandra Katz. Knowledge co-construction and initiative in peer learning interactions. In *AIED 2009, The 14th International Conference on Artificial Intelligence in Education*, Brighton, UK, July 2009.

[228] Jong W. Kim, Frank E. Ritter, and Richard J. Koubek. An integrated theory for improved skill acquisition and retention in the three stages of learning. *Theoretical Issues in Ergonomics Science*, 14(1):22–37, 2013.

[229] Jung Hee Kim, Michael Glass, and Martha W. Evens. Learning use of discourse markers in tutorial dialogue for an intelligent tutoring system. In *COGSCI 2000, Proceedings of the 22nd Annual Meeting of the Cognitive Science Society*, Philadelphia, PA, 2000.

[230] Päivi Kinnunen and Beth Simon. CS majors' self-efficacy perceptions in CS1: results in light of social cognitive theory. In *Proceedings of the seventh International Workshop on Computing Education Research*, pages 19–26. ACM, 2011.

[231] Kenneth R. Koedinger, Vincent Aleven, and Neil T. Heffernan. Toward a rapid development environment for cognitive tutors. In *12th Annual Conference on Behavior Representation in Modeling and Simulation*, Scottsdale, Arizona, USA, 2003.

[232] Kenneth R. Koedinger, John R. Anderson, William H. Hadley, and Mary Mark. Intelligent tutoring goes to school in the big city. *International Journal of Artificial Intelligence in Education*, 8:30–43, 1997.

[233] Boicho Kokinov and Robert M. French. Computational models of analogy-making. In Lynn Nadel, editor, *Encyclopedia of Cognitive Science*, volume 1, pages 113–118. Nature Publishing Group London, 2003.

[234] Georgios Kostopoulos, Anastasia-Dimitra Lipitakis, Sotiris Kotsiantis, and George Gravvanis. Predicting student performance in distance higher education using active learning. In *International Conference on*

Engineering Applications of Neural Networks, pages 75–86. Springer, 2017.

[235] Robert Kozma. The material features of multiple representations and their cognitive and social affordances for science understanding. *Learning and Instruction*, 13:205–226, 2003.

[236] Klaus Krippendorff. *Content Analysis: an Introduction to its Methodology*. Sage Publications, Beverly Hills, CA, 1980.

[237] Klaus Krippendorff. Reliability in Content Analysis. Some Common Misconceptions and Recommendations. *Human Communication Research*, 30(3):411–433, July 2004.

[238] Abhinav Kumar, Barbara Di Eugenio, Jillian Aurisano, and Andrew Johnson. Augmenting Small Data to Classify Contextualized Dialogue Acts for Exploratory Visualization. In *LREC2020, Proceedings of the 12th Language Resources and Evaluation Conference*, pages 590–599, 2020.

[239] Amruth N. Kumar. Model-based reasoning for domain modeling, explanation generation and animation in an ITS to help students learn C++. In *ITS-02 Workshop on Model-based Systems and Qualitative Reasoning for Intelligent Tutoring Systems*, 2002.

[240] Amruth N. Kumar. Generation of problems, answers, grade, and feedback—case study of a fully automated tutor. *ACM Journal of Educational Resources in Computing*, 5(3), September 2005.

[241] Amruth N. Kumar. The effect of using problem-solving software tutors on the self-confidence of female students. In *Proceedings of SIGSCE '08*, pages 523–527, Portland, Oregon, 2008.

[242] Amruth N. Kumar. Results from repeated evaluation of an online tutor on introductory computer science. In *Frontiers in Education Conference (FIE)*, Rapid City, SD, IEEE, 2011.

[243] Amruth N. Kumar. A study of stereotype threat in Computer Science. In *Proceedings of the 17th ACM Annual Conference on Innovation and Technology in Computer Science Education*, Haifa, Israel, pages 273–278, 2012.

[244] Amruth N. Kumar. An evaluation of self-explanation in a programming tutor. In *ITS 2014, Proceedings of the 11th International Conference on Intelligent Tutoring Systems*, pages 248–253. Springer, 2014.

[245] Amruth N. Kumar. The effect of using online tutors on the self-efficacy of learners. In *2015 IEEE Frontiers in Education Conference (FIE)*, El Paso, TX, USA, pages 1–7, 2015.

[246] Amruth N. Kumar. Providing the option to skip feedback in a worked example tutor. In *ITS 2016, Proceedings of the 13th International Conference on Intelligent Tutoring Systems*, pages 101–110. Springer, 2016.

[247] Amruth N. Kumar. Providing the option to skip feedback–a reproducibility study. In *ITS 2019, Proceedings of the 16th International Conference on Intelligent Tutoring Systems*, pages 180–185. Springer, 2019.

[248] Amruth N. Kumar. Long term retention of programming concepts learned using a software tutor. In *ITS 2020, Proceedings of the 17th International Conference on Intelligent Tutoring Systems*, pages 382–387. Springer, 2020.

[249] Amruth N. Kumar and Lisa C. Kaczmarczyk. Programming tutors, practiced concepts, and demographics. In *2013 IEEE Frontiers in Education Conference (FIE)*, Oklahoma, OK, USA, pages 773–778. IEEE, 2013.

[250] Harshit Kumar, Arvind Agarwal, Riddhiman Dasgupta, and Sachindra Joshi. Dialogue act sequence labeling using hierarchical encoder with CRF. In *Thirty-Second AAAI Conference on Artificial Intelligence*, New Orleans, LA, USA, 2018.

[251] Dragan Lambić, Biljana Dorić, and Saša Ivakić. Investigating the effect of the use of Code.org on younger elementary school students' attitudes towards programming. *Behaviour & Information Technology*, pages 1–12, 2020.

[252] Thomas K. Landauer and Susan T. Dumais. A solution to Plato's problem: The Latent Semantic Analysis theory of acquisition, induction, and representation of knowledge. *Psychological Review*, 104:211–240, 1997.

[253] H. Chad Lane and Kurt VanLehn. Coached program planning: Dialogue-based support for novice program design. In *Proceedings of the Thirty-Fourth Technical Symposium on Computer Science Education (SIGCSE '03)*, pages 148–152. ACM Press, 2003.

[254] H. Chad Lane and Kurt VanLehn. Teaching the tacit knowledge of programming to novices with natural language tutoring. *Computer Science Education*, 15(3):183–201, 2005. Special issue on doctoral research in CS Education.

[255] Pat Langley. A general theory of discrimination learning. In D. Klahr, P. Langley, and R. Neches, editors, *Production system models of learning and development*, pages 99–161. The MIT Press, Cambridge, MA, 1987.

[256] Annabel Latham, Keeley Crockett, and David McLean. An adaptation algorithm for an intelligent natural language tutoring system. *Computers & Education*, 71:97–110, 2014.

[257] Annabel Latham, Keeley Crockett, David McLean, and Bruce Edmonds. A conversational Intelligent Tutoring System to automatically predict learning styles. *Computers & Education*, 59(1):95–109, 2012. CAL 2011.

[258] Yann LeCun, Yoshua Bengio, and Geoffrey Hinton. Deep learning. *Nature*, 521(7553):436–444, 2015.

[259] Mark R. Lepper, Michael F. Drake, and Teresa O'Donnell-Johnson. Scaffolding techniques of expert human tutors. In K. Hogan and M. Pressley, editors, *Scaffolding student learning: Instructional approaches and issues*. Brookline, Cambridge, MA, 1997.

[260] Stephen Levinson. *Pragmatics*. Cambridge Textbook in Linguistics. Cambridge University Press, 1983.

[261] Colleen M. Lewis. Exploring variation in students' correct traces of linear recursion. In *Proceedings of the Tenth Annual Conference on International Computing Education Research*, ICER '14, pages 67–74, New York, NY, USA, 2014. ACM.

[262] Colleen M. Lewis. Twelve tips for creating a culture that supports all students in computing. *ACM Inroads*, 8(4):17–20, 2017.

[263] Xiang Li, Hongbo Zhang, Yuanxin Ouyang, Xiong Zhang, and Wenge Rong. A Shallow BERT-CNN Model for Sentiment Analysis on MOOCs Comments. In *2019 IEEE International Conference on Engineering, Technology and Education (TALE)*, Yogyakarta, Indonesia, pages 1–6. IEEE, 2019.

[264] Raymond Lister and John Leaney. Introductory programming, criterion-referencing, and Bloom. In *SIGCSE'03: Proceedings of the 34th SIGCSE Technical Symposium on Computer Science Education*, pages 143–147, New York, NY, USA, 2003. ACM Press.

[265] Diane J. Litman and Kate Forbes-Riley. Predicting student emotions in computer-human tutoring dialogues. In *Proceedings of the 42nd Annual Meeting on Association for Computational Linguistics*, pages 351–359. Association for Computational Linguistics, Baltimore, MD, USA, 2004.

[266] Diane J. Litman and Kate Forbes-Riley. Correlations between dialogue acts and learning in spoken tutoring dialogues. *Natural Language Engineering*, 12(2):161–176, 2006.

[267] Diane J. Litman, Carolyn Penstein Rosé, Kate Forbes-Riley, Kurt VanLehn, Dumisizwe Bhembe, and Scott Silliman. Spoken versus typed human and computer dialogue tutoring. *International Journal of Artificial Intelligence in Education*, 16:145–170, 2006.

[268] Ran Liu and Kenneth R. Koedinger. Closing the loop: Automated data-driven cognitive model discoveries lead to improved instruction and learning gains. *Journal of Educational Data Mining*, 9(1):25–41, 2017.

[269] Zhongxiu Liu, Behrooz Mostafavi, and Tiffany Barnes. Combining Worked Examples and Problem Solving in a Data-Driven Logic Tutor. In *ITS 2016, Proceedings of the 17th International Conference on Intelligent Tutoring Systems*, pages 347–353. Springer, 2016.

[270] Evelyn Lulis, Reva K. Freedman, and Martha Evens. Implementing analogies using APE rules in an electronic tutoring system. In C.K. Looi, G. McCalla, and H. Pain, editors, *AIED 05, the 12th International Conference on AI in Education*, pages 866–888. IOS Press, Amsterdam, 2005.

[271] Evelyn Lulis, Joel Michael, and Martha Evens. How human tutors employ analogy to facilitate understanding. In *Twenty Sixth Annual Meeting of the Cognitive Science Society*, Chicago IL, 2004.

[272] Brian MacWhinney. *The CHILDES project. Tools for analyzing talk: Transcription Format and Programs*, volume 1. Lawrence Erlbaum, Mahwah, NJ, third edition, 2000.

[273] Nabin Maharjan and Vasile Rus. A tutorial Markov analysis of effective human tutorial sessions. In *Proceedings of the 5th Workshop on Natural Language Processing Techniques for Educational Applications*, pages 30–34, Melbourne, Australia, July 2018. Association for Computational Linguistics.

[274] Anshuman Majumdar. The hour of code: An initiative to break the barriers of coding. *XRDS: Crossroads, The ACM Magazine for Students*, 24(3):12–13, 2018.

[275] Mehdi Malekzadeh, Mumtaz Begum Mustafa, and Adel Lahsasna. A review of emotion regulation in Intelligent Tutoring Systems. *Journal of Educational Technology & Society*, 18(4):435–445, 2015.

[276] Ye Mao, Rui Zhi, Farzaneh Khoshnevisan, Thomas W. Price, Tiffany Barnes, and Min Chi. One minute is enough: Early prediction of student success and event-level difficulty during novice programming tasks. In *EDM 2019, Proceedings of the 12th International Conference on Educational Data Mining*, Montréal, Canada, 2019.

[277] Gary Marcus and Ernest Davis. *Rebooting AI: Building Artificial Intelligence we can Trust*. Vintage, 2019.

[278] Jane Margolis, Rachel Estrella, Joanna Goode, Jennifer Jellison Holme, and Kim Nao. *Stuck in the Shallow End: Education, Race, and Computing*. MIT Press, 2nd edition, 2017.

[279] Jane Margolis, Jennifer Jellison Holme, Rachel Estrella, Joanna Goode, Kim Nao, and Simeon Stumme. The Computer Science pipeline in urban high schools: access to what? for whom? *IEEE Technology and Society Magazine*, 22(3):12–19, 2003.

[280] Jane Margolis, Jean J. Ryoo, Cueponcaxochitl D.M. Sandoval, Clifford Lee, Joanna Goode, and Gail Chapman. Beyond access: Broadening participation in high school computer science. *ACM Inroads*, 3(4):72–78, 2012.

[281] Jeffrey D. Marx and Karen Cummings. Normalized change. *American Journal of Physics*, 75(1):87–91, 2007.

[282] Santosh A. Mathan and Kenneth R. Koedinger. Fostering the intelligent novice: Learning from errors with metacognitive tutoring. *Educational Psychologist*, 40(4):257–265, 2005.

[283] Jessica McBroom, Kalina Yacef, and Irena Koprinska. A hierarchical clustering algorithm for trends in periodic educational data. In *AIED'20, The 21st International Conference on Artificial Intelligence in Education*, Ifrane, Morocco, July 2020.

[284] John McCarthy. *LISP 1.5 programmer's manual*. The MIT Press, 1965.

[285] Renee McCauley, Brian Hanks, Sue Fitzgerald, and Laurie Murphy. Recursion vs. iteration: An empirical study of comprehension revisited. In *Proceedings of the 46th ACM Technical Symposium on Computer Science Education*, pages 350–355, 2015.

[286] Lara McConnaughey. An analysis of introductory courses affect on student sentiment and stereotype toward computer science. Master's project UCB/EECS-2019-96, University of California Berkeley, May 2019.

[287] Daniel D. McCracken. Ruminations on Computer Science curricula. *Communications of the ACM*, 30(1):3–5, 1987.

[288] Steven McGee, Ronald I. Greenberg, Lucia Dettori, Andrew M. Rasmussen, Randi McGee-Tekula, Jennifer Duck, and Erica Wheeler. An examination of factors correlating with course failure in a high school Computer Science course. Technical Report 5, The Learning Partnership, 2018. Retrieved from Loyola eCommons, Computer Science: Faculty Publications and Other Works.

[289] Steven McGee, Randi McGee-Tekula, Jennifer Duck, Lucia Dettori, Don Yanek, Andrew Rasmussen, Ronald I. Greenberg, and Dale F. Reed. Does Exploring Computer Science Increase Computer Science Enrollment. In *American Education Research Association Annual Meeting*, volume 1, New York, NY, April 2018.

[290] Andrew McGettrick, Eric Roberts, Daniel D. Garcia, and Chris Stevenson. Rediscovering the passion, beauty, joy and awe: making computing fun again. In *Proceedings of the 39th ACM Technical Symposium on Computer Science Education*, SIGCSE '08, pages 217–218. ACM, 2008.

[291] Jean McKendree. Effective feedback content for tutoring complex skills. *Human-Computer Interaction*, 5:381–413, 1990.

[292] Bruce M. McLaren, Sung-Joo Lim, and Kenneth R. Koedinger. When and how often should worked examples be given to students? New results and a summary of the current state of research. In *Proceedings of the 30th Annual Conference of the Cognitive Science Society*, Washington, DC, USA, pages 2176–2181, 2008.

[293] Bruce M. McLaren, Tamara van Gog, Craig Ganoe, Michael Karabinos, and David Yaron. The efficiency of worked examples compared to erroneous examples, tutored problem solving, and problem solving in computer-based learning environments. *Computers in Human Behavior*, 55:87–99, 2016.

[294] George Herbert Mead. *Mind, Self, and Society*, volume 111. University of Chicago Press, 1934.

[295] Jerry Mead, Simon Gray, John Hamer, Richard James, Juha Sorva, Caroline St. Clair, and Lynda Thomas. A cognitive approach to identifying measurable milestones for programming skill acquisition. In *ITiCSE-WGR '06: Working group reports on Innovation and Technology in Computer Science Education*, pages 182–194, New York, NY, USA, 2006. ACM.

[296] David E. Meltzer. Relation between students' problem-solving performance and representational format. *American Journal of Physics*, 73(5):463–478, May 2005.

[297] Joel A. Michael and Allen A. Rovick. *Problem-solving in physiology*. Prentice Hall, Upper Saddle River, NJ, 1999.

[298] George A. Miller. The magical number seven, plus or minus two. *Psychological Review*, 63(2): 81–97, 1956.

[299] George A. Miller, Eugene Galanter, and Karl H. Pribram. *Plans and the structure of behavior*. Holt, Rinehart and Winston Inc., New York, 1960.

[300] Antonija Mitrović. An intelligent SQL tutor on the web. *International Journal of Artificial Intelligence in Education*, 13(2–4):173–197, 2003.

[301] Antonija Mitrović and Stellan Ohlsson. Evaluation of a constraint-based tutor for a data-base language. *International Journal of Artificial Intelligence in Education*, 10:238–256, 1999.

[302] Antonija Mitrović, Stellan Ohlsson, and Devon K. Barrow. The effect of positive feedback in a constraint-based Intelligent Tutoring System. *Computers & Education*, 60(1):264–272, 2013.

[303] Antonija Mitrović, Stellan Ohlsson, and Brent Martin. Problem-solving support in constraint-based tutors. *Technology, Instruction, Cognition and Learning*, 3(1), 2006.

[304] Behrooz Mostafavi and Tiffany Barnes. Evolution of an intelligent deductive logic tutor using data-driven elements. *International Journal of Artificial Intelligence in Education*, 27(1):5–36, 2017.

[305] Behrooz Mostafavi, Guojing Zhou, Collin Lynch, Min Chi, and Tiffany Barnes. Data-driven worked examples improve retention and completion in a logic tutor. In *International Conference on Artificial Intelligence in Education*, pages 726–729. Springer, 2015.

[306] Dejana Mullins, Anne Deiglmayr, and Hans Spada. Motivation and emotion in shaping knowledge co–construction. In Michael Baker, Sanna Jarvela, and Jerry Andriessen, editors, *Affective Learning Together: Social and Emotional Dimensions of Collaborative Learning*, pages 139–161. Routledge, 2013.

[307] Laurie Murphy, Sue Fitzgerald, Scott Grissom, and Renée McCauley. Bug infestation! A goal-plan analysis of CS2 students' recursive binary tree solutions. In *Proceedings of the 46th ACM Technical Symposium on Computer Science Education*, Kansas City, MO, USA, pages 482–487, 2015.

[308] Tom Murray, Klaus Schultz, David Brown, and John Clement. An analogy-based computer tutor for remediating physics misconceptions. *Interactive Learning Environments*, 1(2):79–101, 1990.

[309] Amir Shareghi Najar and Antonija Mitrović. Do novices and advanced students benefit differently from worked examples and ITS? In *Int. Conf. Computers in Education*, Jakarta, Indonesia, pages 20–29, 2013.

[310] Amir Shareghi Najar, Antonija Mitrović, and Bruce M. McLaren. Adaptive Support versus Alternating Worked Examples and Tutored Problems: Which Leads to Better Learning? In Dimitrova V., Kuflik T., Chin D., Ricci F., Dolog P., and Houben G.J., editors, *User Modeling, Adaptation, and Personalization. UMAP 2014*, pages 171–182. Springer, 2014.

[311] Amir Shareghi Najar, Antonija Mitrović, and Bruce M. McLaren. Examples and tutored problems: adaptive support using assistance scores. In *Proceedings of the 24th International Conference on Artificial Intelligence*, pages 4317–4323. AAAI Press, 2015.

[312] Amir Shareghi Najar, Antonija Mitrović, and Bruce M. McLaren. Learning with intelligent tutors and worked examples: selecting learning activities adaptively leads to better learning outcomes than a fixed curriculum. *User Modeling and User-Adapted Interaction*, 26(5):459–491, 2016.

[313] National Academies of Sciences, Engineering, and Medicine. *Assessing and Responding to the Growth of Computer Science Undergraduate Enrollments*. National Academies Press, 2018.

[314] National Research Council. *Report of a Workshop on the Scope and Nature of Computational Thinking*. National Academies Press, 2010.

[315] David M. Neves and John R. Anderson. Knowledge compilation: Mechanisms for the automatization of cognitive skill. In John R. Anderson, editor, *Cognitive Skills and their Acquisition*, pages 57–84. Hillsdale, NJ: Erlbaum, 1981.

[316] Allen Newell and Paul S. Rosenbloom. Mechanisms of skill acquisition and the law of practice. In John R. Anderson, editor, *Cognitive Skills and their Acquisition*, pages 1–55. Lawrence Erlbaum, Hillsdale, NJ, 1981.

[317] Allen Newell, John Calman Shaw, and Herbert A. Simon. Empirical explorations of the logic theory machine: a case study in heuristic. In *Western Joint Computer Conference: Techniques for Reliability*, pages 218–230. ACM, Los Angeles, CA, USA, 1957.

[318] Allen Newell and Fred M. Tonge. An introduction to Information Processing Language V. *Communications of the ACM*, 3(4):205–211, 1960.

[319] Timothy J. Nokes and Kurt VanLehn. Bridging principles and examples through analogy and explanation. In *Proceedings of the 8th International Conference on International Conference for the Learning Sciences-Volume 3*, pages 100–102. International Society of the Learning Sciences, Utrecht, The Netherlands, 2008.

[320] Elnaz Nouri and David R. Traum. Initiative Taking in Negotiation. In *15th Annual Meeting of the ACL Special Interest Group on Discourse and Dialogue*, pages 186–193, Philadelphia, PA, USA, 2014.

[321] Benjamin D. Nye, Arthur C. Graesser, and Xiangen Hu. AutoTutor and family: A review of 17 years of natural language tutoring. *International Journal of Artificial Intelligence in Education*, 24(4):427–469, 2014.

[322] Stellan Ohlsson. Truth versus appropriateness: Relating declarative to procedural knowledge. In David Klahr, Pat Langley, and Robert T. Neches, editors, *Production System Models of Learning and Development*, pages 287–327. The MIT Press, Cambridge, MA, 1987.

[323] Stellan Ohlsson. Constraint-based student modeling. *Journal of Artificial Intelligence and Education*, 3(4):429–447, 1992. Reprinted in J. E. Greer and G. I. McCalla (Eds.), *Student Modeling: The Key to Individualized Knowledge-Based Instruction*, Springer Verlag, 1994.

[324] Stellan Ohlsson. Learning from error and the design of task environments. *International Journal of Educational Research*, 25(5):419–448, 1996.

[325] Stellan Ohlsson. Learning from performance errors. *Psychological Review*, 103:241–262, 1996.

[326] Stellan Ohlsson. Learning by specialization and order effects in the acquisition of cognitive skills. In E. Ritter, J. Nerb, T. O'Shea, and E. Lehtinen, editors, *In Order to Learn: How the Sequence of Topics Affects Learning*. Oxford University Press, New York, NY, 2007.

[327] Stellan Ohlsson. Computational models of skill acquisition. In R. Sun, editor, *The Cambridge Handbook of Computational Psychology*, pages 359–395, Cambridge, UK, 2008. Cambridge University Press.

[328] Stellan Ohlsson. *Deep Learning: How the Mind Overrides Experience*. Cambridge University Press, 2011.

[329] Stellan Ohlsson, Barbara Di Eugenio, Bettina Chow, Davide Fossati, Xin Lu, and Trina C. Kershaw. Beyond the code-and-count analysis of tutoring dialogues. In *AIED 2007, the 13th International Conference on Artificial Intelligence in Education*, Marina del Rey, CA, July 2007.

[330] Stellan Ohlsson and Antonjia Mitrović. Fidelity and efficiency of knowledge representations for Intelligent Tutoring Systems. *Technology, Instruction, Cognition and Learning*, 5:101–132, 2007.

[331] José Paladines and Jaime Ramírez. An Intelligent Tutoring System for Procedural Training with Natural Language Interaction. In *CSEDU 2019, the International Conference on Computer Supported Education,*, Heraklion, Crete, Greece, May 2019.

[332] Seymour Papert. *Mindstorms: Children, Computers, and Powerful Ideas*. Basic Books, Inc., 1980.

[333] Sagar Parihar, Ziyaan Dadachanji, Praveen Kumar Singh, Rajdeep Das, Amey Karkare, and Arnab Bhattacharya. Automatic grading and feedback using program repair for introductory programming courses. In *Proceedings of the 2017 ACM Conference on Innovation and Technology in Computer Science Education*, pages 92–97, 2017.

[334] Rebecca J. Passonneau and Diane J. Litman. Discourse segmentation by human and automated means. *Computational Linguistics*, 23(1):103–139, 1997.

[335] Jeffrey Pennington, Richard Socher, and Christopher D. Manning. Glove: Global vectors for word representation. In *Proceedings of the 2014 Conference on Empirical Methods in Natural Language Processing (EMNLP)*, Doha, Qatar, pages 1532–1543, 2014.

[336] Natalie K. Person, Arthur C. Graesser, Joseph P. Magliano, and Roger J. Kreuz. Inferring what the student knows in one-to-one tutoring: The role of student questions and answers. *Learning and Individual Differences*, 6(2):205–229, 1994.

[337] Jean Piaget. *The Psychology of Intelligence*. Routledge, London, 1950.

[338] Peter L. Pirolli and John R. Anderson. The role of learning from examples in the acquisition of recursive programming skills. *Canadian Journal of Psychology*, 39(2):240–272, 1985.

[339] Peter L. Pirolli and John R. Anderson. The role of learning from examples in the acquisition of recursive programming skills. *Canadian Journal of Psychology/Revue canadienne de psychologie*, 39(2):240, 1985.

[340] Peter L. Pirolli and Margaret Recker. Learning strategies and transfer in the domain of programming. *Cognition and Instruction*, 12(3):235–275, 1994.

[341] Heather Pon-Barry, Brady Clark, Karl Schultz, Elizabeth Owen Bratt, and Stanley Peters. Advantages of spoken language interaction in dialogue-based Intelligent Tutoring Systems. In *International Conference on Intelligent Tutoring Systems*, pages 390–400. Springer, 2004.

[342] Leo Porter, Mark Guzdial, Charlie McDowell, and Beth Simon. Success in introductory programming: What works? *Communications of the ACM*, 56(8):34–36, 2013.

[343] Kiki Prottsman. Computer Science for the Elementary Classroom. *ACM Inroads*, 5(4):60–63, December 2014.

[344] Ruixiang Qi and Davide Fossati. Unlimited trace tutor: Learning code tracing with automatically generated programs. In *SIGCSE 2020, the 51st ACM Technical Symposium on Computer Science Education*, Portland, OR, March 2020.

[345] Ross J. Quinlan. *C4.5: Programs for Machine Learning*. Morgan Kaufmann, Los Altos, CA, 1993.

[346] Colin Raffel, Noam Shazeer, Adam Roberts, Katherine Lee, Sharan Narang, Michael Matena, Yanqi Zhou, Wei Li, and Peter J. Liu. Exploring the limits of transfer learning with a unified text-to-text transformer. *Journal of Machine Learning Research*, 21(140):1–67, 2020.

[347] Jamie Raigoza. A study of students' progress through introductory Computer Science programming courses. In *2017 IEEE Frontiers in Education Conference (FIE)*, pages 1–7, 2017.

[348] Chaitanya Ramineni and Paul Deane. The Criterion® online writing evaluation service. In Scott A. Crossley and Danielle S. McNamara, editors, *Adaptive Educational Technologies for Literacy Instruction*. Taylor & Francis, Routledge: NY, 2016.

[349] Brian J. Reiser, John R. Anderson, and Robert G. Farrell. Dynamic Student Modelling in an Intelligent Tutor for LISP Programming. In *IJCAI 1985, Proceedings of the Ninth International Joint Conference on Artificial Intelligence*, Los Angeles, CA, USA, pages 8–14, 1985.

[350] Ehud Reiter. An architecture for data-to-text systems. In *ENLG07, Proceedings of the 11th European Workshop on Natural Language Generation*, Saarbruecken, Germany, June 2007.

[351] Alexander Renkl. The worked-out-example principle in multimedia learning. *The Cambridge Handbook of Multimedia Learning*, pages 229–245, 2005.

[352] V.G. Renumol, Dharanipragada Janakiram, and S. Jayaprakash. Identification of cognitive processes of effective and ineffective students during computer programming. *Trans. Comput. Educ.*, 10(3):1–21, 2010.

[353] Lauren B. Resnick, John M. Levine, and Stephanie D. Teasley. *Perspectives on Socially Shared Cognition*. American Psychological Association, Washington, DC, 1991.

[354] Debra J. Richardson. Student-focused initiatives for retaining students in computing programs. *ACM Inroads*, 9(2):13–18, 2018.

[355] Timothy C. Rickard. Bending the power law: A CMPL theory of strategy shifts and the automatization of cognitive skills. *Journal of Experimental Psychology: General*, 126(3):288–311, 1997.

[356] Toni Rietveld and Roeland van Hout. *Statistical Techniques for the Study of Language and Language Behaviour*. Mouton de Gruyter, Berlin - New York, 1993.

[357] Christian Rinderknecht. A survey on teaching and learning recursive programming. *Informatics in Education-An International Journal*, 13(1):87–120, 2014.

[358] Frank E. Ritter, Farnaz Tehranchi, and Jacob D. Oury. ACT-R: A cognitive architecture for modeling cognition. *Wiley Interdisciplinary Reviews: Cognitive Science*, 10(3), 2019.

[359] Kelly Rivers and Kenneth R. Koedinger. Data-driven hint generation in vast solution spaces: a self-improving Python programming tutor. *International Journal of Artificial Intelligence in Education*, 27(1):37–64, 2017.

[360] Michele Roberts, Kiki Prottsman, and Jeff Gray. Priming the Pump: Reflections on Training K-5 Teachers in Computer Science. In *Proceedings of the 49th ACM Technical Symposium on Computer Science Education*, SIGCSE '18, pages 723–728, New York, NY, USA, 2018. Association for Computing Machinery.

[361] Carolyn Penstein Rosé, Pam Goldman, Jennifer Zoltners Sherer, and Lauren Resnick. Supportive technologies for group discussion in MOOCs. *Current Issues in Emerging eLearning*, 2(1), 2015. Article 5.

[362] Carolyn Penstein Rosé, Pamela W. Jordan, Michael Ringenberg, Stephanie Siler, Kurt VanLehn, and Anders Weinstein. Interactive conceptual tutoring in Atlas-Andes. In *AI-ED 2001, Proceedings of the 10th International Conference on Artificial Intelligence in Education*, San Antonio, TX, USA, 2001.

[363] Carolyn Penstein Rosé, Elizabeth A. McLaughlin, Ran Liu, and Kenneth R. Koedinger. Explanatory learner models: Why machine learning (alone) is not the answer. *British Journal of Educational Technology*, 50(6):2943–2958, 2019.

[364] Carolyn Penstein Rosé, Johanna D. Moore, Kurt VanLehn, and David Albritton. A comparative evaluation of socratic versus didactic tutoring. In *23rd Annual Conference of the Cognitive Science Society*, Edinburgh, Scotland, August 2001.

[365] Paul S. Rosenbloom and Allen Newell. Learning by chunking: A production system model of practice. In David Klahr, Pat Langley, and Robert T. Neches, editors, *Production System Models of Learning and Development*, pages 221–286. The MIT Press, Cambridge, MA, 1987.

[366] Miguel Angel Rubio. Automated prediction of novice programmer performance using programming trajectories. In *AIED 2020, Proceedings of the 21st International Conference on Artificial Intelligence in Education*, Ifrane, Morocco, 6-10 July 2020.

[367] Vasile Rus, Peter Brusilovsky, Scott Fleming, Lasang Tamang, Kamil Akhuseyinoglu, Jordan Barria-Pineda, Nisrine Ait-Khayi, and Zeyad Alshaikh. An Intelligent Tutoring System for Source Code Comprehension. In *AIED 2019, Proceedings of the 20th International Conference on Artificial Intelligence in Education*, Chicago, IL, USA, June 25-29 2019.

[368] Vasile Rus, Nabin Maharjan, and Rajendra Banjade. Dialogue act classification in human-to-human tutorial dialogues. In *International Conference on Smart Learning Environments (ICSLE)*, pages 185–188. Springer, Beijing, China, 2017.

[369] Leonardo Sa and Wen-Jung Hsin. Traceable recursion with graphical illustration for novice programmers. *InSight: A Journal of Scholarly Teaching*, 5:54–62, 2010.

[370] Swarnadeep Saha, Tejas I. Dhamecha, Smit Marvaniya, Renuka Sindhgatta, and Bikram Sengupta. Sentence level or token level features for automatic short answer grading?: Use both. In *AIED 2019, Proceedings of the 19th International Conference on Artificial Intelligence in Education*, pages 503–517. Springer, 2018.

[371] Dario D. Salvucci and John R. Anderson. Integrating analogical mapping and general problem solving: the path-mapping theory. *Cognitive Science*, 25:67–110, 2001.

[372] Gregory A. Sanders, Martha W. Evens, Gregory D. Hume, Allen A. Rovick, and Joel A. Michael. An analysis of how students take the initiative in keyboard-to-keyboard tutorial dialogues in a fixed domain. In *Proceedings of the Fourteenth Annual Conference of the Cognitive Science Society*, Bloomington, IN, USA, pages 1086–1091, 1992.

[373] Ian Sanders, Vashti Galpin, and Tina Götschi. Mental models of recursion revisited. *SIGCSE Bull.*, 38(3):138–142, June 2006.

[374] Kay G. Schulze, Robert N. Shelby, Donald J. Treacy, and Mary C. Wintersgill. Andes: A Coached Learning Environment for Classical Newtonian Physics. In *Proceedings of the 11th International Conference on College Teaching and Learning*, Jacksonville, FL, April 2000.

[375] A. Scime. Globalized computing education: Europe and the United States. *Computer Science Education*, 18(1):43–64, 2008.

[376] Allison Scott, Alexis Martin, Frieda McAlear, and Tia C. Madkins. Broadening Participation in Computer Science: Existing Out-of-School Initiatives and a Case Study. *ACM Inroads*, 7(4):84–90, November 2016.

[377] Terry Scott. Bloom's taxonomy applied to testing in Computer Science classes. *Journal of Computing Sciences in Colleges*, 19(1):267–274, 2003.

[378] John R. Searle. What Is a Speech Act. In Max Black, editor, *Philosophy in America*, pages 615–628. Cornell University Press, Ithaca, New York, 1965. Reprinted in *Pragmatics. A Reader*, Steven Davis editor, Oxford University Press, 1991.

[379] John R. Searle. Indirect Speech Acts. In P. Cole and J.L. Morgan, editors, *Syntax and Semantics 3. Speech Acts.* Academic Press, 1975. Reprinted in *Pragmatics. A Reader*, Steven Davis editor, Oxford University Press, 1991.

[380] Burr Settles. *Active Learning*, volume 6 of *Synthesis Lectures on Artificial Intelligence and Machine Learning*. Morgan & Claypool Publishers, 2012.

[381] Jai Seu, Ru-Charn Chang, Jun Li, Martha W. Evens, Joel A. Michael, and Allen A. Rovick. Language difference in face-to-face and keyboard-to-keyboard tutoring sessions. In *Proceedings of the Thirteenth Annual Conference of the Cognitive Science Society*, Chicago, IL, USA, pages 576–580, 1991.

[382] Farhana Shah, Martha W. Evens, and Joel A. Michael. Classifying student initiatives and tutor responses in human keyboard-to-keyboard tutoring sessions. *Discourse Processes*, 33(1):23–52, 2002.

[383] Harsh Shah and Amruth N. Kumar. A tutoring system for parameter passing in programming languages. *ACM SIGCSE Bulletin*, 34(3):170–174, 2002.

[384] Razieh Sheikhpour, Mehdi Agha Sarram, Sajjad Gharaghani, and Mohammad Ali Zare Chahooki. A survey on semi-supervised feature selection methods. *Pattern Recognition*, 64:141–158, 2017.

[385] Elizabeth Shriberg, Raj Dhillon, Sonali Bhagat, Jeremy Ang, and Hannah Carvey. The ICSI Meeting Recorder Dialog Act (MRDA) Corpus. In *Proceedings of the 5th SIGdial Workshop on Discourse and Dialogue at HLT-NAACL 2004*, pages 97–100, 2004.

[386] Valerie J. Shute. Focus on formative feedback. *Review of Educational Research*, 78(1):153–189, 2008.

[387] Satinder Singh, Diane J. Litman, Michael Kearns, and Marilyn Walker. Optimizing Dialogue Management with Reinforcement Learning: Experiments with the NJFun System. *Journal of Artificial Intelligence Research (JAIR)*, 15:105–133, 2002.

[388] Rebecca Smith and Scott Rixner. The Error Landscape: Characterizing the Mistakes of Novice Programmers. In *Proceedings of the 50th ACM Technical Symposium on Computer Science Education*, SIGCSE '19, pages 538–544, New York, NY, USA, 2019. Association for Computing Machinery.

[389] Lawrence Snyder, Tiffany Barnes, Daniel D. Garcia, Jody Paul, and Beth Simon. The first five Computer Science Principles pilots: summary and comparisons. *ACM Inroads*, 3(2):54–57, 2012.

[390] Elliot Soloway, Beth Adelson, and Kate Ehrlich. Knowledge and processes in the comprehension of computer programs. In Michelene T.H. Chi, Robert Glaser, and Marshall J. Farr, editors, *The nature of expertise*, pages 129–152. Lawrence Erlbaum Associates Hillsdale, NJ, 1988.

[391] Elliott Soloway and James C. Spohrer. *Studying the Novice Programmer*. Lawrence Erlbaum Associates, Mahwah, NJ, USA, 1988.

[392] Robert A. Sottilare, Arthur Graesser, Xiangen Hu, and Heather Holden. *Design Recommendations for Intelligent Tutoring Systems*, volume 1: Learner Modeling. US Army Research Laboratory, 2013.

[393] John Stasko, Albert Badre, and Clayton Lewis. Do algorithm animations assist learning?: an empirical study and analysis. In *Proceedings of the INTERACT'93 and CHI'93 Conferences on Human Factors in Computing Systems*, pages 61–66. ACM, 1993.

[394] Andreas Stolcke, Klaus Ries, Noah Coccaro, Elisabeth Shriberg, Rebecca Bates, Daniel Jurafsky, Paul Taylor, Rachel Martin, Carol Van Ess-Dykema, and Marie Meteer. Dialogue act modeling for automatic tagging and recognition of conversational speech. *Computational Linguistics*, 26(3):339–373, 2000.

[395] Emma Strubell, Ananya Ganesh, and Andrew McCallum. Energy and policy considerations for deep learning in NLP. In *Proceedings of the 57th Annual Meeting of the Association for Computational Linguistics*, pages 3645–3650, Florence, Italy, July 2019. Association for Computational Linguistics.

[396] Ron Sun, Edward Merrill, and Todd Peterson. From implicit skills to explicit knowledge: A bottom-up model of skill learning. *Cognitive Science*, 25(2):203–244, 2001.

[397] Pramuditha Suraweera and Antonija Mitrović. An intelligent tutoring system for entity relationship modeling. *International Journal of Artificial Intelligence in Education*, 14(3–4):375–417, 2004.

[398] Vanessa Svihla, Woong Lim, Elizabeth Ellen Esterly, Irene A. Lee, Melanie E. Moses, Paige Prescott, and Tryphenia B. Peele-Eady. Designing for assets of diverse students enrolled in a freshman-level Computer Science For All course. In *ASEE Annual Conference Proceedings*, Columbus, Oh, USA, 2017.

[399] John Sweller. Cognitive load during problem solving: Effects on learning. *Cognitive Science*, 12(2):257–285, 1988.

[400] John Sweller and Graham A. Cooper. The use of worked examples as a substitute for problem solving in learning algebra. *Cognition and Instruction*, 2(1):59–89, 1985.

[401] Lasang Jimba Tamang, Zeyad Alshaikh, Nisrine Ait Khayi, and Vasile Rus. The effects of open self-explanation prompting during source code comprehension. In *The Thirty-Third International FLAIRS Conference*, Online, pages 451–456, 2020.

[402] Betty Tärning. Review of Feedback in Digital Applications–Does the Feedback They Provide Support Learning? *Journal of Information Technology Education: Research*, 17(1):247–283, 2018.

[403] Maryam Tavafi, Yashar Mehdad, Shafiq Joty, Giuseppe Carenini, and Raymond Ng. Dialogue act recognition in synchronous and asynchronous conversations. In *Proceedings of the SIGDIAL 2013 Conference*, Metz, France, pages 117–121, 2013.

[404] Josh Tenenberg and Laurie Murphy. Knowing what I know: An investigation of undergraduate knowledge and self-knowledge of data structures. *Computer Science Education*, 15(4):297–315, 2005.

[405] Caitlin Tenison, Jon M. Fincham, and John R. Anderson. Phases of learning: How skill acquisition impacts cognitive processing. *Cognitive Psychology*, 87:1–28, 2016.

[406] Joe Tessler, Bradley Beth, and Calvin Lin. Using Cargo-Bot to Provide Contextualized Learning of Recursion. In *Proceedings of the 9th Annual International ACM Conference on Computing Education Research*, pages 161–168. ACM, San Diego, CA, USA, 2013.

[407] Thomas James Tiam-Lee and Kaoru Sumi. Analysis and prediction of student emotions while doing programming exercises. In *International Conference on Intelligent Tutoring Systems*, pages 24–33. Springer, 2019.

[408] Lam Tik. Effectiveness of Hour of Code to Enhance Computational Thinking of Hong Kong Junior Secondary Students-Echo with STEM Education in Hong Kong. Technical report, The Education University of Hong Kong, 2017. Undergraduate Project.

[409] David R. Traum. 20 questions on dialogue act taxonomies. *Journal of Semantics*, 17(1):7–30, 2000.

[410] Sho-Huan Tung, Ching-Tao Chang, Wing-Kwong Wong, and Jihn-Chang Jehng. Visual representations for recursion. *International Journal of Human-Computer Studies*, 54(3):285–300, 2001.

[411] Franklyn Turbak, Constance Royden, Jennifer Stephan, and Jean Herbst. Teaching recursion before loops in CS1. *Journal of Computing in Small Colleges*, 14(4):86–101, 1999.

[412] Anna C. K. van Duijvenvoorde, Kiki Zanolie, Serge A. R. B. Rombouts, Maartje E. J. Raijmakers, and Eveline A. Crone. Evaluating the Negative or Valuing the Positive? Neural Mechanisms Supporting Feedback-Based Learning across Development. *Journal of Neuroscience*, 28(38):9495–9503, 2008.

[413] Kurt VanLehn. Rule acquisition events in the discovery of problem-solving strategies. *Cognitive Science*, 15:1–47, 1991.

[414] Kurt VanLehn. Rule learning events in the acquisition of a complex skill: An evaluation of cascade. *Journal of the Learning Sciences*, 8(1):71–125, 1999.

[415] Kurt VanLehn. The behavior of tutoring systems. *International Journal of Artificial Intelligence in Education*, 16(3):227–265, 2006.

[416] Kurt VanLehn, Winslow Burleson, Sylvie Girard, Maria Elena Chavez-Echeagaray, Javier Gonzalez-Sanchez, Yoalli Hidalgo-Pontet, and Lishan Zhang. The affective meta-tutoring project: lessons learned. In *International Conference on Intelligent Tutoring Systems*, pages 84–93. Springer, 2014.

[417] Kurt VanLehn, Reva Freedman, Pamela W. Jordan, Charles Murray, Remus Osan, Michael Ringenberg, Carolyn Penstein Rosé, Kay Schultze, Robert Shelby, Donald Treacy, Anders Weinstein, and Mary Wintersgill. Fading and deepening: The next steps for ANDES and other model-tracing tutors. In *ITS 2000, Proceedings of the 5th International Conference on Intelligent Tutoring Systems*, Montreal, Canada, 2000.

[418] Kurt VanLehn, Arthur C. Graesser, G. Tanner Jackson, Pamela W. Jordan, Andrew Olney, and Carolyn Penstein Rosé. When are tutorial dialogues more effective than reading? *Cognitive Science*, 31(1):3–62, 2007.

[419] Kurt VanLehn, Pamela W. Jordan, and Carolyn Penstein Rosé. The architecture of Why2-Atlas: A coach for qualitative physics essay writing. In S.A. Cerri, G. Gouarderes, and F. Paraguacu, editors, *ITS 2002, the 6th International Conference on Intelligent Tutoring Systems*, pages 158–167, 2002.

[420] Elsa Q. Villa. Minority voices: Interrupting the social environment to retain undergraduates in computing. *ACM Inroads*, 9(3):31–33, August 2018.

[421] Stella Vosniadou and Anthony Ortony, editors. *Similarity and analogical reasoning*. Cambridge University Press, 1989.

[422] Lev S. Vygotsky. *Mind and Society: The Development of Higher Mental Processes*. Harvard University Press, Cambridge, MA, USA, 1978.

[423] Henry M. Walker. Retention of Students in Introductory Computing Courses: Curricular Issues and Approaches. *ACM Inroads*, 8(4):14–16, 2017.

[424] Marilyn A. Walker, Rebecca Passonneau, and Julie E. Boland. Quantitative and qualitative evaluation of DARPA Communicator spoken dialogue systems. In *Proceedings of the 39th Annual Meeting on Association for Computational Linguistics*, pages 515–522. Association for Computational Linguistics, 2001.

[425] Marilyn A. Walker and Steve Whittaker. Mixed initiative in dialogue: an investigation into discourse segmentation. In *Proceedings of the 28th Annual Meeting of the Association for Computational Linguistics*, pages 70–78, Morristown, NJ, USA, 1990. Association for Computational Linguistics.

[426] Tsung-Hsien Wen, David Vandyke, Nikola Mrkšić, Milica Gašić, Lina M. Rojas-Barahona, Pei-Hao Su, Stefan Ultes, and Steve Young. A network-based end-to-end trainable task-oriented dialogue system. In *Proceedings of the 15th Conference of the European Chapter of the Association for Computational Linguistics: Volume 1, Long Papers*, pages 438–449, Valencia, Spain, April 2017. Association for Computational Linguistics.

[427] John Wieting, Jonathan Mallinson, and Kevin Gimpel. Learning paraphrastic sentence embeddings from back-translated bitext. In *Proceedings of the 2017 Conference on Empirical Methods in Natural Language Processing*, Copenhagen, Denmark, pages 274–285, 2017.

[428] Joseph B. Wiggins, Kristy Elizabeth Boyer, Alok Baikadi, Aysu Ezen-Can, Joseph F. Grafsgaard, Eun Young Ha, James C. Lester, Christopher M. Mitchell, and Eric N. Wiebe. JavaTutor: an Intelligent Tutoring System that adapts to cognitive and affective states during computer programming (Demo). In *Proceedings of the 46th ACM Technical Symposium on Computer Science Education*, Kansas City, MO, USA, page 599, 2015.

[429] Joseph B. Wiggins, Joseph F. Grafsgaard, Kristy Elizabeth Boyer, Eric N. Wiebe, and James C. Lester. Do you think you can? the influence of student self-efficacy on the effectiveness of tutorial dialogue for computer science. *International Journal of Artificial Intelligence in Education*, 27(1):130–153, Mar 2017.

[430] Derek Wilcocks and Ian Sanders. Animating recursion as an aid to instruction. *Computers & Education*, 23(3):221–226, 1994.

[431] Kevin H. Wilson, Xiaolu Xiong, Mohammad Khajah, Robert V. Lindsey, Siyuan Zhao, Yan Karklin, Eric G. Van Inwegen, Bojian Han, Chaitanya

Ekanadham, Joseph E. Beck, Neil Heffernan, and Michael C. Mozer. Estimating student proficiency: Deep learning is not the panacea. In *In Neural Information Processing Systems, Workshop on Machine Learning for Education*, Barcelona, Spain, page 3, 2016.

[432] Jeannette M. Wing. Computational thinking. *Communications of the ACM*, 49(3):33–35, May 2006.

[433] Jeannette M. Wing. Computational thinking and thinking about computing. *Philosophical Transactions of the Royal Society*, 366:3717–3725, July 2008.

[434] Leon E. Winslow. Programming pedagogy—A psychological overview. *SIGCSE Bull.*, 28(3):17–22, 1996.

[435] Ian H. Witten, Eibe Frank, Mark A. Hall, and Christopher J. Pal. *Data Mining: Practical Machine Learning Tools and Techniques*. Morgan Kaufmann, 2016.

[436] Cheng-Chih Wu, Nell B. Dale, and Lowell J. Bethel. Conceptual models and cognitive learning styles in teaching recursion. *SIGCSE Bull.*, 30(1):292–296, March 1998.

[437] Wanli Xing and Dongping Du. Dropout prediction in MOOCs: Using deep learning for personalized intervention. *Journal of Educational Computing Research*, 57(3):547–570, 2019.

[438] Fan Yang and Peter A. Heeman. Initiative conflicts in task-oriented dialogue. *Computer Speech & Language*, 24(2):175–189, 2010.

[439] Kai-Hsiang Yang and Hui-Ying Lin. Exploring the Effectiveness of Learning Scratch Programming with Code.org. In *2019 8th International Congress on Advanced Applied Informatics (IIAI-AAI)*, pages 1057–1058, 2019.

[440] Christina Zeller and Ute Schmid. Automatic generation of analogous problems to help resolving misconceptions in an Intelligent Tutor System for written subtraction. Technical report, University of Bamberg, 2017.

[441] Haoran Zhang, Ahmed Magooda, Diane J. Litman, Richard Correnti, Elaine Wang, Lindsay C. Matsumura, Emily Howe, and Rafael Quintana. eRevise: Using Natural Language Processing to provide formative feedback on text evidence usage in student writing. In *Proceedings of the AAAI Conference on Artificial Intelligence*, volume 33, pages 9619–9625, 2019.

[442] Lishan Zhang, Yuwei Huang, Xi Yang, Shengquan Yu, and Fuzhen Zhuang. An automatic short-answer grading model for semi-open-ended questions. *Interactive Learning Environments*, pages 1–14, 2019.

[443] Daniel Zingaro, Cynthia Taylor, Leo Porter, Michael Clancy, Cynthia Lee, Soohyun Nam Liao, and Kevin C. Webb. Identifying student difficulties with basic data structures. In *Proceedings of the 2018 ACM Conference on International Computing Education Research*, pages 169–177, 2018.

[444] Stuart Zweben and Betsy Bizot. 2019 CRA Taulbee Survey. *Computing Research News*, 32(5), May 2019.